China's Petroleum Industry

Chu-yuan Cheng

The Praeger Special Studies program—utilizing the most modern and efficient book production techniques and a selective worldwide distribution network—makes available to the academic, government, and business communities significant, timely research in U.S. and international economic, social, and political development.

China's Petroleum Industry

Output Growth and Export Potential

PRAEGER SPECIAL STUDIES IN INTERNATIONAL ECONOMICS AND DEVELOPMENT

205013

Praeger Publishers   New York   Washington   London

Library of Congress Cataloging in Publication Data

Cheng, Chu-yuan.
    China's petroleum industry.

    (Praeger special studies in international economics and
development)
    Bibliography: p.
    Includes index.
    1. Petroleum industry and trade—China.   I. Title.
HD9576.C52C49   338.2'7'2820951   75-19771
ISBN 0-275-01810-5

PRAEGER PUBLISHERS
111 Fourth Avenue, New York, N.Y. 10003, U.S.A.

Published in the United States of America in 1976
by Praeger Publishers, Inc.

Printed in the United States of America

To the memory of my parents

Mr. Cheng Hung-shan (1886-1970)

Mrs. Yang Shu-chen (1891-1975)

Ever since the opening of the Tach'ing oil field in North Manchuria 15 years ago, the Chinese petroleum industry has received the limelight from her national press, and in recent years, as a consequence of the energy crisis in the West and subsequent claims of the discovery of offshore reserves in the continental shelf of the Yellow and South China Seas, the Chinese petroleum industry has attracted worldwide attention. Paralleling this attention has been a flood of widely divergent conjectures about China's crude oil output and its future export potential. Unfortunately, these free-flowing "guesstimates" are largely undocumented and afford little in the way of systematic analysis of the admittedly scant and fragmentary data available. The purpose of this study is to provide a thorough and systematic analysis of this largely unexplored data, focusing on the growth of output and the pattern of development of the Chinese petroleum industry from the year 1949 and projecting its output, consumption, and potential for export for the 1975-85 time frame.

This study is addressed to the following four sets of related questions:

1. How reliable are official claims with respect to crude oil output and newly discovered reserves? What is China's potential for becoming a major oil producer?

2. Since the petroleum industry is both capital-intensive and technology-intensive, can China develop her petroleum industry without the cooperation of Japan and the West?

3. Assuming China's petroleum industry continues to grow rapidly, will China shift from primary dependence upon coal as an energy source to primary dependence upon oil?

4. Between 1975 and 1985, will the petroleum industry serve as catalyst by stimulating the modernization of China's agriculture and industry?

In order to attempt to answer the above questions, this study will first examine the output goals of the Chinese development plans for various time frames since 1949. These goals will then be measured against official output data and cross-checked by aggregation of the production data from the individual oilfields, derived by microanalysis. The study next focuses on oil reserves, refining, and transportation facilities. It also surveys China's capacity for manufacturing petroleum extracting and refining equipment and probes China's procurement of foreign equipment and machinery. Based upon both quantitative and qualitative information, the aggregate demand and the aggregate supply of petroleum products for the 1949-85 period are estimated, as well as China's potential for exporting oil.

To assess the contribution of the petroleum industry to China's general industrialization and modernization, reliance is placed on its relative share of the value of gross industrial output, employment, and balance of payments and the overall linkage effects generated by its petroleum industry.

This study relies heavily on official Chinese data; as is typical of the statistical data made available by the People's Republic, the data on the petroleum industry are both scant and fragmentary. Limited statistics on crude oil output and technical reports on the geological conditions of oil fields in the northwest region are available for the 1952-60 period, but since 1961 very little official data on output or capacity has been released. To bridge these data gaps for the post-1961 period, a wide variety of official publications have been extensively studied. Many major local newspapers and printed copies of local radio programs were also screened. Fragmentary though they are, these are the best available sources of up-to-date information on the Chinese petroleum industry.

The quest for data has also led to extensive scrutinizing of secondary sources published in Taiwan, Hong Kong, and Japan. Japanese reports published by various government agencies, business firms, and trade associations also contain a modicum of current information on Chinese petroleum production. Although lacking documentation and sometimes conflicting, these sources are very valuable for cross-checking other sources. For the pre-1960 era, Russian-language materials have also been very informative.

Based on these diverse sources of information, some 50 statistical tables were constructed. Obviously, many of the data gaps can be filled only by arbitrary conjectures and interpolations. To test the internal consistency of these derivations and projections, a series of cross-checks has been made. For example, not only is the total crude oil output per year compared with the aggregated output of the major oil fields, but the output data is also compared with consumption, imports, and exports. The production figures are further checked with refinery capacity for the same period. The total crude oil output for given periods (1952-62, 1962-72) is then compared with the estimated capital investment to calculate the capital-output ratio for these periods. No claim is made for the precision of these estimates, except that they are based on the best data available.

The study was financed by the Social Science Research Council and subsequently by the Trade Analysis Division, Bureau of East-West

Trade, Department of Commerce. I am indebted to Professors
Albert Feuerwerker and Robert Dernberger of the University of
Michigan, Dwight Perkins of Harvard University, K. C. Yeh of Rand
Corporation, and John Gurley of Stanford University, members of
the Subcommittee on Research on the Chinese Economy of the Social
Science Research Council, for the support of this project. I am also
grateful to Dr. Allen J. Lenz, Dr. William Clarke, Mr. Albert
Jankowitz, Dr. Nai-Ruenn Chen, and Dr. David Denny, all of the
Bureau of East-West Trade, for their assistance and support. To
Professors Alexander Eckstein of the University of Michigan;
Richard Burkhardt, Joseph Black, John Hannaford, and Maurice
Girgis of Ball State; and Ta-Chung Liu of Cornell University I ex-
tend sincere thanks for their constant encouragement and concern
for my research.

Special thanks also to Mr. Kikuzo Ito of the Tokyo Shimbun,
who supplied me with many useful Japanese reports; to Mr. Wei-ying
Wan and Mr. Wei-yi Ma of the Asia Library at the University of
Michigan; and to Dr. Chi Wang and Mr. P. K. Tseng of the Chinese
Section of the Library of Congress for helping to obtain the Chinese
materials. Dr. Laurence Ma kindly lent me his thesis on industrial
development in Northwestern China.

I wish also to extend my appreciation to Joel Caron for his
critical review of the first three chapters of the manuscript. Robert
Jost painstakingly edited the entire manuscript: not only did he significantly improve its presentability and readability, but he also
clarified many ambiguities by asking searching questions and pin-
pointing a number of pitfalls, and to him I am indebted deeply. Kay
Pilkington and Laura Townsend typed the entire manuscript, and
Susan Nelson helped to draw the maps and figures; to all of them I
extend my thanks.

Finally, I must express my personal debt and apologies to my
wife Hua, who has assumed the entire burden of family chores,
proofread my manuscript, and yet struggled forward with her own
dissertation.

CONTENTS

LIST OF TABLES, FIGURES, AND MAPS

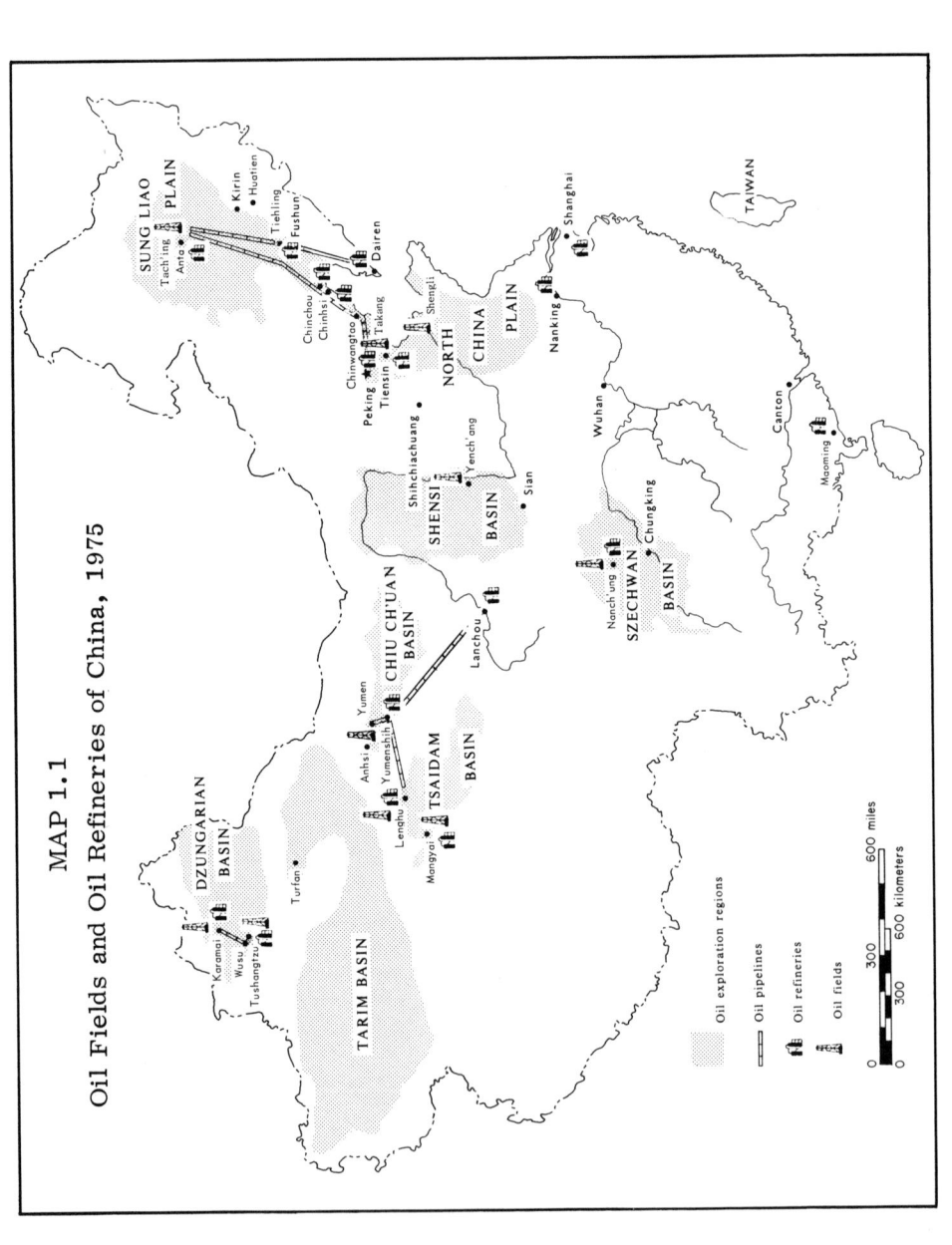

MAP 1.1

Oil Fields and Oil Refineries of China, 1975

SUNG LIAO PLAIN
Tach'ing
Anta
Kirin
Huaten
Tiehling
Fushun
Chinchou
Chinhsi
Chinwangtao
Peking
Tiensin
Tangku
Shengli
Dairen
NORTH CHINA PLAIN
Shanghai
Nanking
Shihchiachuang
Yench'ang
Sian
SHENSI BASIN
Wuhan
Canton
Maoming
Chungking
Nanch'ung
SZECHWAN BASIN
CHIU CH'UAN BASIN
Lanchou
Yumen
Anhsi
Yumenshih
Lenghu
Mongyai
TSAIDAM BASIN
Turfan
DZUNGARIAN BASIN
Karamai
Wusu
Tushangtzu
TARIM BASIN
TAIWAN

Oil exploration regions

Oil pipelines

Oil refineries

Oil fields

300   600 miles

0   300   600 kilometers

China's Petroleum Industry

FIGURE 1.1

Crude Oil Production in Mainland China, 1950-74

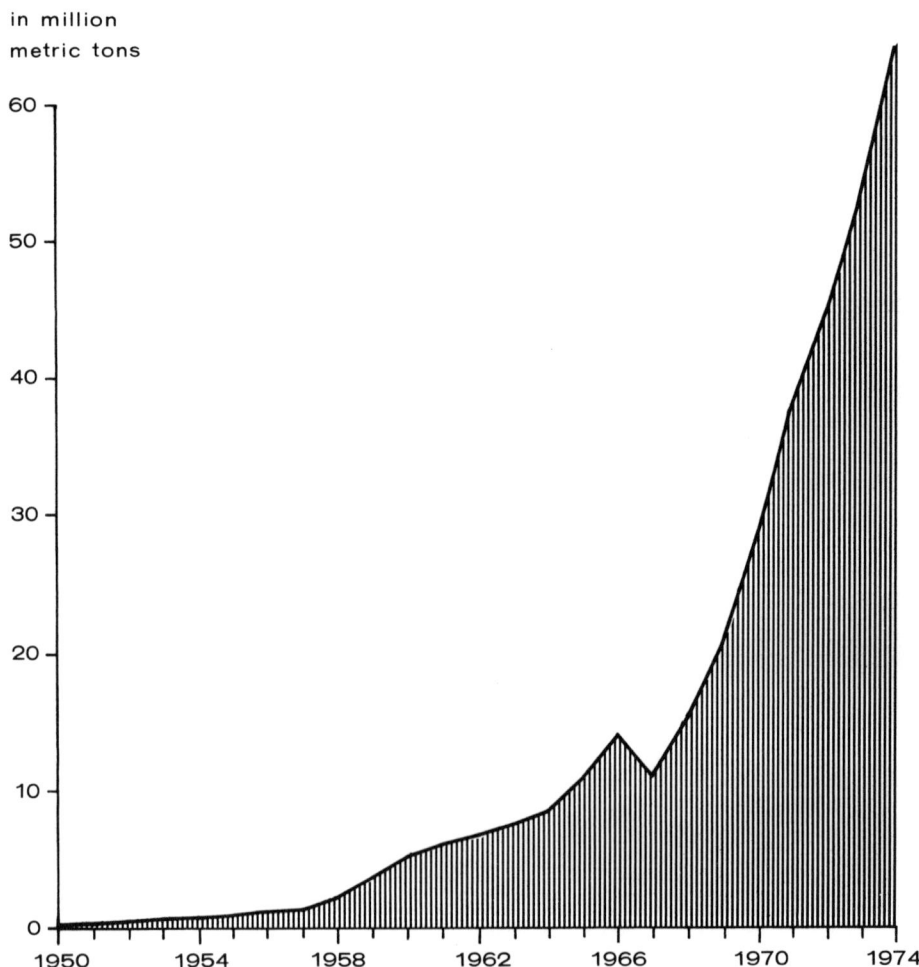

1

THE DEVELOPMENT OF THE CHINESE PETROLEUM INDUSTRY

Since the mid-1960s the most potent industrial development in the People's Republic of China has been the metamorphosis of the petroleum industry, which has garnered recognition for China as a potential major exporter of crude oil.

As of 1957, Chinese planners considered the petroleum industry the weakest link in the national economy: output was 27.5 percent short of fulfilling the First Five-Year Plan goal and provided only one-third of domestic consumption.

Five additional years of gestation prefaced an exuberant growth rate after 1963, a consequence of better development strategy and technological change. The achievement of self-sufficiency in petroleum products by 1968 demonstrated the impact of this phenomenal transformation and showed China's potential for becoming a major petroleum exporter.

The objective of this chapter is to trace and assess the consequences of past developmental policies and technological changes on China's petroleum industry.

HISTORICAL BACKGROUND

China was among the first countries of the world to discover petroleum. Records of oil are found in the History of the Han Dynasty (Han Hsu), written 1,800 years ago by the famous historian Pan Ku. In 1878 the Manchu government decided to drill for oil at Miaoli in Taiwan. Two engineers from the United States, using American equipment, commenced exploratory operations, but an accident occurred while they were drilling the shaft that precluded further activity. In 1907 the first productive oil well on the Chinese mainland was opened in Yen-ch'ang, located in Shensi Province. However, because of shoddy exploratory work and poor assessment of the geological formation, the work was stopped in the belief that large-scale exploration was not profitable.[1] In the 19 years after the exploration in Yen-ch'ang, attempts to extract oil in China were abortive. China's modern petroleum industry dates from the discovery of the Yumen oil field in Kansu in 1936. In 1940, with Russian aid, the Wusu and Tushantsu oil fields in Sinkiang began operation.

During Japan's occupation of Manchuria in the 1930s, synthetic oil plants were built in Fushun by extracting oil from the abundant oil shale available in that area. Oil refineries were built in Fushun, Chinchou, Kirin, and Szeping. Total crude oil production in 1943, the peak of the pre-1949 era, was 318,000 tons, most of which (255,000 tons) came from Fushun.[2]

In the northwest frontier province of Sinkiang, where the Russian influence was more prevalent, a more elaborate geological survey was conducted by the Russians in the 1940s, resulting in the discovery of two promising oil fields: one in the north of the T'ienshan from Tihua (Urumchi) to Tach'eng; the other in the south of the T'ienshan from Kuch'e to Kashgar. Crude oil reserves were estimated at 258,833,000 tons and oil shale at 500 million tons.[3]

In August 1945, shortly before the end of the Sino-Japanese war, the Soviet government declared war on Japan, and after the Japanese surrender immediately dispatched armed forces to Manchuria. The petroleum industry in Manchuria suffered extensive destruction, since the Soviets removed most of the principal facilities. The synthetic oil plants became paralyzed, and total output of crude oil in 1948 dropped to 86,000 tons.

From the inception of the Communist government in 1949 to the end of 1952, prior to the First Five-Year Plan, the Chinese petroleum industry underwent a rapid rehabilitation. Production facilities in Yumen in the northwest and in Fushun in Manchuria were restored. In 1950 a Sino-Soviet joint stockholding company for petroleum refining was established in Sinkiang to explore the Tushantsu oil field. Five years later the formation of the Ministry for Geology brought all of the country's geological work under unified direction. Some 79,651 meters of oil wells were drilled between 1949 and 1952. By 1952 crude oil production reached 436,000 tons, an increase of 260 percent over 1949 and 36.1 percent over 1943.

THE FIRST FIVE-YEAR PLAN

The further expansion of production capacity and the discovery of additional new major oil fields paralleled the implementation of

the First Five-Year Plan in 1953. Since the Five-Year Plan emphasized large-scale industrialization, the demand for oil increased
rapidly. Industrial demands pointed up the weakness of the technological base of this infant industry; moreover, the conditions of the
mineral resources were largely unknown. Western geologists had
long maintained that both the type of rocks and their genetic age precluded the possibility of there being any petroleum deposits worthy
of exploitation throughout most of China.[4] For example, Standard
Oil geologists M. L. Fuller and F. G. Clapp, who were sent to explore petroleum resources in China in 1913-15, concluded that "A
large part of the Chinese Republic consists of rocks of type and age
in which no possibility of commercial oil deposits exists."[5] Eliot
Blackwelder of Stanford University declared that "China will never
produce large quantities of oil."[6] This theory had a stifling effect
on the Chinese effort to develop the petroleum industry.

The First Five-Year Plan was not published until 1955. It
specified concentrated efforts to prospect natural oil resources, to
increase production of synthetic oil, and to expand the entire petroleum industry in the long run. The initial investment for these projects amounted to 702 million yuan or 3 percent of the total capital investment for industry during the five-year period. The plan outlined
extensive geological surveying and drilling in Kansu, Sinkiang,
Szechwan, and Tsinghai provinces and encouraged the maximum en-
largement of proven reserves. Plan fulfillment by 1957 stipulated
(1) an increase of 730 percent over the oil drilling footage of 1952
and (2) proven oil reserves of 55.18 million tons--an increase of
280 percent over 1952.

To expand production, the plan provided for the construction
of 13 petroleum projects, two of which would be designed with the
help of the Soviet Union and nine of which would be completed during
the five-year period. Production capacity was to increase by
1,520,000 tons and refinery capacity by 1.5 times that of 1952.[7]

To achieve these petroleum production goals and to emphasize
their importance, a Ministry of Petroleum Industry was established
in July 1955. A Bureau of Petroleum Geology was also formulated
under the Ministry of Geology. The developmental policy during this
period apparently followed the basic Russian exploration philosophy
of "shoot for the elephants and ignore the mice."[8] The government's
primary effort focused on large-scale exploration in major basins--
the Dzungaria, Tsaidam, Turfan, Pre-nan Shan, Sungliao, and
Szechwan. Increasing crude oil output became a secondary object.
In preparing for large-scale prospecting, geological workers had to
be recruited and trained. In 1949 it was reported that only 18 petroleum geologists with eight drilling machines were engaged in oil exploration; by 1955 there were over 6,000 geological workers, with

several hundred drilling machines at their disposal.[9] A nationwide prospecting campaign was conducted in 1956. Total footage of drilling reached 404,612 meters, which was 2.07 times that of 1955 and 5.5 times that of 1950.[10]

As a result of this extensive prospecting, within an area of half a million square kilometers more than 300 oil structures and 240 oil seepages were discovered. It was declared that oil possibilities were promising in the Tsaidam Basin of Tsinghai, the Dzungaria Basin of Sinkiang, the Chiu-ch'uan Basin of Kansu, the Szechwan area, the Shensi Basin (Ordos Area), the Kweichow area, the Tibet area, and the north China plains.[11]

The First Five-Year Plan produced mixed results. Capital investment greatly surpassed the limits of the original plan. Actual investment totaled 1.9 billion yuan, which was almost three times the original plan and accounted for 8 percent of actual investment in modern industry.[12]

The extensive prospecting led to the opening of three new oil structures in Yumen and the discoveries of the Karamai and Tsaidam oil fields. During these five years, drilling footage in Yumen was 26 times greater than during the 11 years before 1949; crude oil output amounted to five times the total for those 11 years.

The capacity of refinery and synthetic oil plants was also greatly expanded. Of these plants, the Lanchow Refinery was the most prominent. Built with modern Soviet equipment, it acquired an annual processing capacity of one million tons of crude oil. It was not completed, however, until 1960.

Despite these successes, the original goals of the plan were not attained. The 1957 crude oil output reached only 1.46 million tons--72.5 percent of the goal.

The failure to achieve the production targets can be attributed to several factors. First, the developmental policy in this period placed high priority on prospecting of natural oil resources rather than on production. It was correctly believed that systematic prospecting was a necessary prerequisite for future large-scale production. For this reason, half of the crude oil produced in this period came from synthetic oil plants that had been established in the 1930s. From a technological standpoint, however, numerous problems in hydrogenation and distillation remained unsolved, which prevented a rapid growth of synthetic oil production. Since only one ton of oil was being extracted from 30 tons of shale, the shale oil production was not considered economical.[13]

Second, the petroleum industry is characterized by a long gestation period. The exploration of a large oil field takes three to four years. Investment in exploration in Karamai, Tsaidam, and Yumen during this period did not yield significant results until after 1958.

For instance, oil was detected at the end of 1955 in Kramai and drilling began in 1956, but actual production started in 1958.[14] In Tsidam,
prospecting for the principal oil field at Lenghu started in 1955 and
the first gusher was drilled in 1958.[15]

Third, most of the laborers and technicians recruited had been
soldiers, farmers, or herdsmen, who were ill-prepared to handle
modern equipment and unfamiliar with industrial discipline. This
resulted in an extremely high accident rate and very low productivity.
During the first half of 1957 there were 896 accidents. Time lost because of these accidents accounted for 23.3 percent of the total drilling time for the drilling teams under the Sinkiang Petroleum Administration, 24 percent for those under the Tsinghai Petroleum Administration, and 26.7 percent for those under the Yumen Petroleum
Administration.[16]

The high overhead costs and low productivity led to an ex-
traordinarily high capital-output ratio. According to K. C. Yeh's
estimate, the capital requirement amounted to 480 yuan per ton of
crude oil.* Converted into dollars at the official exchange rate at
that time, the figure in yuans amounted to $205 per ton. The aver-
age U.S. crude oil replacement cost (including exploration cost, development cost, and operation cost) for the same period, 1952-57,
based on 1958 prices was about $19 per ton.[17] Therefore, the
Chinese capital requirement was about ten times that of the United
States.

The failure to attain the target goals had a significant impact
on long-term developmental policies. In September 1956, one year
before the termination of the First Five-Year Plan, a Second Five-
Year Plan (1958-62) was proposed by the Chinese Communist Party's
Eighth Party Congress. In this proposal the output target of crude
oil for 1962 was set at 5 to 6 million tons, or 2.48 to 2.98 times
the planned target for 1957, representing an average annual growth
rate of 20 to 24 percent. However, as the actual output in 1957
reached only 1,460,000 tons instead of the planned 2,012,000, the
1962 planned target was actually 3.4 to 4.1 times that of the 1957

----

*This estimate is derived on the assumption of a two-year time
lag between investment and output. Investment in the petroleum industry in 1953-57 is given at 1.9 billion yuan. Planned investment
in 1958 is reported at 0.34 billion yuan. Total investment in 1953-
58 therefore amounts to 2.24 billion yuan. The increase in output
in 1954-60 is about 4.7 million tons. The capital-output relation is
thus calculated at 480 yuan per ton. (K. C. Yeh, Communist China's
Petroleum Situations (Santa Monica, Calif.: Rand Corporation, 1962),
p. 19.

output and required an average annual growth rate of 27.8 to 32.7 percent. (See Tables 1.1 and 1.2.) In view of the slow growth in the First Five-Year Plan period, the Chinese authorities believed that attainment of the 1962 target goals required new developmental policies. It is against this background that the original developmental plan for the 1958-62 period was soon discarded and replaced by the Great Leap Forward program.

THE GREAT LEAP FORWARD

The general idea behind the Great Leap Forward was simultaneous acceleration of the growth of both the modern sector and the traditional sector of the economy. A drastic drive to build millions of small workshops and mines using indigenous methods of production was vigorously carried out throughout the country. Some 60 million peasants were mobilized by the government to build half a million of the so-called "backyard blast furnaces." A similar program was launched in the petroleum industry. At the Nanch'ung conference, held in April 1958 at Nanch'ung, Szechwan, new guidelines were formulated by the Ministry of Petroleum that called for a developmental program "to start from small to large, from shallow to deep (oil wells), to combine foreign and native methods, to plan for comprehensive utilization of resources, and to strive for self-sufficiency (in petroleum products)."[18] The major goal of the Great Leap Forward was to extract as much oil as possible using the most accelerated methods available.

To increase petroleum output, the new developmental policy thus shifted the focus from exploration to production. Production commenced in Karamai and in Tsaidam in 1958. In Tsaidam, the Number 5 oil field in Lenghu, which was the most prolific oil structure at that time, was put into full operation; during the first six months of 1958, approximately 100 wells were drilled in that area. The operations of the Lenghu structure brought an eightfold increase of crude oil for the Tsaidam oil field in 1959.[19] In Karamai, output in 1959 trebled that of 1958, and in the first half of 1960, output trebled again over that of the comparable period in 1959.[20]

Refinery capacity was markedly expanded. The modern Lanchow refinery, the largest in China, was completed in 1960 and had commenced limited production as early as January 1959.[21] Pipelines were laid between Karamai and Tushantzu and between Yumen and Lanchow. Expansion of the Nanking Refinery was completed in June 1960, giving the refinery an annual capacity of around 2 million tons.

TABLE 1.1

Planned and Actual Output of Crude Oil in the
First and Second Five-Year Plans

Year	Crude Oil Output (thousands of tons)	Index 1949 = 100	Index 1952 = 100
1943	318	263	73
1949	121	100	28
1952	436	360	100
1957 planned target	2,012	1,660	461
1957 actual output	1,458	1,210	335
1962 planned target	5,000-6,000	4,130-4,960	1,150-1,370

Sources: 1943, 1949, and 1952: from Chao I-wen, Hsin-chung-kuo ti Kung-yeh (Peking: 1957), pp. 49-50. 1957: planned target from First Five-Year Plan for Development of the National Economy of the People's Republic of China, 1953-1957 (Peking: Foreign Languages Press, 1956), pp. 73-74. 1957: actual output from State Statistical Bureau, Ten Great Years (Peking: Foreign Languages Press, 1960), p. 95. 1962: planned target from Chinese Communist Party, "Proposal of the Second Five-Year Plan," in Eighth National Congress of the Chinese Communist Party in China (Peking: Foreign Languages Press, 1956).

TABLE 1.2

Percentage of Increase and Annual Growth of Output of Crude
Oil in the First and Second Five-Year Plans

	Increase	Annual Growth Rate
1952-57 (planned growth)	361	35.8
1952-57 (actual growth)	235	27.4
1957-62 (planned growth)	148-198	20-24.4
1957 (actual output)-1962 (planned target)	240-310	27.8-32.7

Sources: From Table 1.1.

Perhaps the most dramatic aspect of the Great Leap was the opening of the huge shallow oil reserves and the construction of thousands of small shale oil plants. Before the Nanch'ung conference, any oil reserve with a rate below one ton per day was considered marginal and not worth drilling. The new policy called for large-scale exploration of shallow oil reserves throughout the country. An official estimate at that time cited 1,020 square kilometers of shallow oil-bearing strata in Sinkiang, Tsinghai, Szechwan, Kweichow, Shensi, and Manchuria. [22] The shallow oil formations in Karamai alone were said to have a potential of 34 million tons of crude oil. [23] Under the new program, some .1 million productive shallow wells would be built in Karamai by using simple equipment and native methods. It was anticipated that by 1962 shallow oil output would account for 26 to 40 percent of the total oil production in Karamai. [24]

The Great Leap Forward emphasized also the full utilization of China's abundant coal and shale-oil reserves. Preliminary surveys estimated China's shale-oil reserves in 21 provinces at 50 billion tons and coal reserves at 1,500 billion tons, a large portion of which could be used as the raw material for oil extraction. [25] It was planned that more than 2,000 small (annual rate from 300 to 1,000 tons) and 100 medium-size synthetic plants (annual rate from 1,000 to 10,000 tons) would be built during 1958-59. Based on plans for 23 provinces and cities, the total production of local medium and small synthetic petroleum plants would amount to more than 2 million tons during the second Five-Year Plan period. [26]

The rationale behind the new policy was low capital-output ratio, quick results, simple technology, native facilities, and proximity to consumer. It was assumed that a shale-oil plant capable of producing 1 million tons annually would take four years to construct. The crude production capacity investment per ton per year was almost 800 yuan, yet only three months would be needed to put a shale or coal pile distillation plant into operation that would produce 1,000 tons annually. The total investment was only about 60,000 to 70,000 yuan, or 60 to 70 yuan per ton per year, and it was assumed that the investment could be recovered within a period of two years. Similarly, a coal or shale-oil furnace-type distillation plant producing 10,000 tons annually could be built within eight to nine months; the total investment of 2 million yuan, or 240 yuan per ton, could hopefully be recovered within a period of little more than three years. [27]

Unlike the result of the Great Leap Forward in other industries, in petroleum the program met with some success. As the developmental policy shifted from prospecting to extracting, crude oil output markedly increased. It rose from 1.458 million tons in 1957 to 2.264 million tons in 1958, an increase of 55 percent. Output reached

3.7 million tons in 1959 and 5.5 million tons in 1960: the 1962 annual
production goal of 5 to 6 million tons set by the second Five-Year
Plan was thereby accomplished two years ahead of the scheduled target date. The average annual growth rate of crude oil production
during these three years (1958-60) reached 56 percent, which more
than doubled the annual growth rate of 27.4 percent attained under
the First Five-Year Plan.

Prospecting during this period brought impressive results.
The prospecting working force had expanded to 36,000 people in
1958, when there were 916 petroleum prospecting teams. The total
footage of drilling in 1958-59 exceeded the total of the entire First
Five-Year Plan period by 124.6 percent.[28] The number of wells
drilled in 1958 was four times greater than that of the previous year.
Several major oil fields that would be developed in later years, including Tach'ing, Shengli, and Takang, were first discovered during
this period.

The costs of these bold achievements were relatively high.
Hasty drilling and exploration practices were reflected in the very
high percentage of technically faulty borings.[29] According to one
official report, of the 21 oil wells sunk in Tsaidam in 1958, 20 failed
to reach the predetermined point underground simply because they
failed to meet the technical requirements of vertical drilling.[30]
Moreover, although more than 1,000 small synthetic plants were
constructed, their aggregate output in 1959 reached only 100,000
tons.[31] It was openly admitted that the extraction of oil from coal
by primitive methods was expensive and that the quality of the oil
produced was very unsatisfactory.

In 1961, after three years of the Great Leap Forward program,
the national economy, instead of moving ahead, suffered a severe
setback. The economic crisis was partly the result of three years
of natural calamities and partly of the hasty regimentation of the
rural communes in 1958, which had caused a 25 percent drop in food
crops during the 1959-61 period. The millions of small workshops
and mines without adequate technical design and equipment turned
out millions of unusable tons of steel and other industrial products
that greatly distorted the resource allocation. The crisis was ex-
acerbated by the sudden withdrawal of Soviet technological and economic aid in August 1960 as a consequence of the intensifying Sino-
Soviet dispute.

As the crisis mounted, the Chinese Communist Party convened
its ninth plenary session in January 1961 to evaluate the situation.
The Great Leap Forward program in industrialization was abandoned.
Additionally, the suspension of Soviet supplies of crude oil and
petroleum products forced China to pursue a policy of self-sufficiency
in petroleum at all costs. Since the supply of Soviet transportation

equipment was also curtailed, the expansion of the oil fields in the northwestern region became acutely less feasible. However, at this critical moment the opening of Tach'ing oil field on the eastern coast greatly alleviated the nation's shortage in petroleum.

EXPLORATION OF EASTERN OIL FIELDS

Prior to 1960 the development of natural oil fields was primarily confined to northwestern China. The three major oil fields within this area, Yumen, Karamai, and Tsaidam, were developed in the 1950-60 period. Since these fields were more than 1,500 miles away from the eastern industrial and population centers, their exploration required very heavy overhead costs. Moreover, shipping millions of tons of crude oil from the remote areas to the refineries on the eastern coast taxed the resources of the only railroad serving these two regions. If China were to develop efficient production of these interior oil resources, a massive program of pipeline construction seemed essential, in order to reduce the cost of delivering oil to the industrial coast centers. Since the construction of pipeline involves heavy capital outlays and long construction periods, which the Chinese were seeking to avoid, the principal alternative was the rapid development of oil and gas resources within the coastal areas. The exploitation of the Tach'ing oil field in Manchuria, the Shengli oil field in the lower Yellow River region, and the Takang oil field in the Peking-Tientsin area were manifestations of these considerations.

Chinese petroleum production reached a turning point in 1964, when the Tach'ing oil field, in the Sungliao plain of Manchuria, commenced large-scale operations. The new oil field, located 160 kilometers northwest of Harbin and 65 kilometers northwest of Anta, became the first major oil-producing center in the eastern area. Prospecting had begun in 1955 with the help of Russian geologists and geophysicists. The discovery well, which was in approximately 1,200 meters of lower cretaceous sandstone, was successfully completed on September 10, 1959.[32] The Tach'ing structure was 20 kilometers wide and 50 kilometers long and contained more than 22 reservoir sandstones at depths up to 1,500 meters (5,000 feet). Ultimate recoverable oil was estimated at 86 to 160 million tons, or 628 million barrels to 1.2 billion barrels. The opening of Tach'ing, which went on limited production in 1962-63 and commenced full-scale production in 1964, brought China close to self-sufficiency in petroleum.[33] The entire complex contained five structures with more than 2,000 operating wells attesting to its impressive output. Between 1960 and 1971 the average annual growth rate in crude oil reached 35.2 percent.[34] In 1974 Tach'ing produced more than 20

million tons of crude oil--approximately one-third of the nation's
total output. A modern refinery was built in 1963 with a designed
capacity of 1 million tons annually, but in 1971 it was expanded to
2.5 million tons, and subsequently it was doubled to 5 million tons. [35]

Tach'ing was considered by the Chinese authorities as the most
significant development in Chinese industrialization. In 1964 Mao
Tse-tung called on the entire industry to "learn from Tach'ing." The
success of Tach'ing was significant to Chinese economic development
in three aspects.

First, its large-scale operation came at an opportune time because the Soviet Union intended to use an oil blockade as a weapon in
the Sino-Soviet dispute. The operation of Tach'ing made China self-
sufficient for this crucial product.

Second, Tach'ing exemplified the spirit of self-reliance. The
operation of the new oil field and the construction of the new refinery
were undertaken by Chinese engineers and technicians. The maturity
of the Chinese technological force made exploration in other new oil
fields much more feasible.

Third, Tach'ing represented a new type of industrial center
envisioned by Mao: a combination of industry and agriculture, and
a mingling of urban districts with rural. From the very beginning,
Tach'ing was designed so as not to replicate the typical oil-boom
city of the West. Major emphasis was given to preserving the grass-
lands. The 400,000 inhabitants were drawn from all parts of the
nation and were resettled in 40 townships scattered over the region,
with urban facilities at the center, each township being linked with
two or three satellite village residential areas. The dependents of
the oil field workers were organized into agricultural production
teams to cultivate the lands surrounding the oil wells and produced
grains and vegetables in order to achieve greater self-sufficiency
within the oil-producing area. The official exhortation of Tach'ing
meant in part that it served as an ideal model in which there was no
overconcentration of population and no sharp distinction between industrial and agricultural works. [36] In fact, it was officially reported
that Tach'ing produced 25,000 tons of grain and 30,000 tons of vegetables a year. [37]

In the wake of Tach'ing, Chinese petroleum workers soon explored two other major oil fields on the eastern coast. The Shengli
oil field, which had two sections, was opened in 1962 and commenced
operations in 1964. The north Shengli is situated in the northern
part of Shantung province along both the southern and northern banks
of the Yellow River. The area includes Kuangjao, Pohsing, Lichin,
Pinhsien, Chanhua, Huimin, and Yanghsin, more than 6,000 square
kilometers. Prospecting work began in 1958, and the test drilling
was initiated in 1960. The first successful well was completed in

1962, and by 1964 the field had commenced operations. By the end
of 1966, 16 wells were completed, and the annual yield exceeded
half a million tons the following year. The crude oil was shipped to
Tsingtao in Shantung by rail and then via coastal shipping to Shanghai
for refining.

The second section of the Shengli oil field was located in a nar-
row band extending from Wei-shan Lake to the north of Hsu-chou,
through the provinces of Kiangsi and Anhuei, terminating in the P'an-
yang Lake in Kiangsi Province. The deposits covered an area of
3,500 square kilometers with a producing area of 1,100 square kilometers in the Ho-fei, Ch'uan-chiao, Lu-kiang triangle. In 1967
some 30 wells produced up to 300,000 tons a year in this area.
Crude oil produced here was taken by tanker to refineries in Nanking.[38] Total output of the Shengli oil field was estimated to have
increased to an impressive 10 million tons in 1974.[39]

More significant than the opening of Shengli was the development of the Takang oil field in the northern portion of the Gulf of
Pohai, with Tientsin as its refining center. Prospecting started in
the spring of 1964, and production began in 1967; the field was found
to have extensive thick oil-bearing formations giving high yields of
good-quality crude oil.[40] Since early 1973, prospecting, exploitation, and construction, as well as extraction, have been expanded.[41]
In 1974 its output of crude reached 4 million tons.

The greater part of the Takang oil field lies under the Gulf of
Pohai, an inland sea stretching deep into the mainland of China.
Almost completely enclosed by the Liaotung Peninsula on the north
and the Shantung Peninsula on the south, it joins the Yellow Sea at
the Pohai straits. Like the land surface, the floor of the Pohai Sea
slopes gently from northwest to southeast with an average depth of
almost 20 meters. This offshore oil resource is considered a part
of the continental shelf, which is believed by geologists to be potentially very productive.[42]

The Takang oil field is still in its initial stage of development,
but its growth has been accelerating in recent years. Offshore drilling presents a very promising prospect to the Chinese petroleum industry: until recently China was thought to have perhaps the 12th-
or 13th-largest oil reserve in the world, which was sufficient to
meet its needs for the immediate future but not nearly enough for it
to become a major industrial power; however, the inclusion of offshore resources lying beneath the Yellow Sea and the East China Sea
has notably enhanced China's potential petroleum resources.

The expansion of Tach'ing, Shengli, and Takang in recent
decades has brought great vitality to the Chinese petroleum industry.
The foremost long-term impact is the change in geographic distribution: in the past there was a critical imbalance in the regional distribution of the Chinese petroleum industry. While 80 percent of the

country's industrial output was produced in the coastal areas, more than 90 percent of the oil reserves were in the remote areas of the northwest. Tsaidam, Yumen, and Karamai are all located more than 1,500 miles west of Shanghai. By development of inland and offshore oil resources near the industrialized coastal areas, the costs of production have been significantly reduced. In recent years there has been a significant downward adjustment of prices for petroleum products, including an 18.6 percent reduction in gasoline prices in 1965,[43] a 20.8 percent reduction in the kerosene price in 1971, and a 9.7 percent decline for diesel oil in the same year.[44] Thus, industry and agriculture on the eastern coast enjoy the benefits of lowered petroleum product prices.

SOME PECULIAR FEATURES

Compared with that of other major oil producers in the world, the developmental path traversed by the Chinese petroleum industry since 1949 exhibits several unique features.

A prominent feature of the experience of China is her adherence to the principle of independence and self-reliance. Unlike the oil producers in the Organization of Petroleum Exporting Countries (OPEC), China is far less susceptible to the influence of foreign oil companies. Despite the fact that China relied quite heavily on the Soviet supply of technological assistance and equipment until 1960 and has eagerly sought western technology in recent years, the Chinese petroleum industry is now relatively independent, and official statements of the 1970s stress that China's petroleum industry will develop with a minimum of foreign assistance. It will "move along the road of maintaining independence and keeping the initiative in our own hands."[45] When approached by Japan for joint offshore exploration between Chinese and Japanese producers, Chou En-lai pointedly declined and ruled out the possibility of participation of Japanese enterprises in the development of Chinese oil resources.[46] The Chinese policy differs from that of the USSR, who in recent years has anxiously sought Japanese cooperation in exploiting oil resources in Siberia.

The second unique feature of the Chinese developmental policy is the concern over oil resource conservation. Since World War II, as world consumption of oil tripled, the output of crude oil in most oil-producing countries increased tremendously, resulting in steep depletion of their oil resources. In the United States the ratios between proven reserves and production have fallen steadily, from 16.3 years in 1920 to 14.1 years in 1940, 12.8 years in 1960, and 8.9 years in 1970.[47] From the very beginning the Chinese authorities

considered conservation of resources to be as important as increases
in production. "Increase production while practicing economy" is an
established policy. Between 1965 and 1972, although the Chinese
petroleum output rose threefold, the government never loosened its
grip on consumption control. Moreover, the Chinese leaders have
categorically stated that "China will not become a supplier of resources, nor will China allow foreign countries to acquire China's
resources." The economic wisdom of such policies is controversial,
but if adhered to, they may tend to slow the depletion of oil resources
in China.

The third unique feature of China's developmental policy is the
way she developed her oil fields. Confronted with critical shortages
of modern equipment and technological manpower, the Chinese leaders tended to apply their old guerrilla warfare techniques to developing oil fields. After deciding to explore the Tach'ing oil field in
February 1960, Mao ordered an "annihilation campaign to open up
the field"--his old military strategy of concentrating an overwhelm-
ing force at the point of attack. More than 80,000 workers, technicians, and soldiers converged from all directions. It was reported
that 75.5 percent of the equipment and 80 percent of the workers in
the northwest Yumen oil fields were transferred to other oil fields,
primarily to support Tach'ing.[48] Later, in developing other eastern
oil fields, large numbers of workers and drilling machines left
Tach'ing. Recent Western sources report that some 36,000 workers
are now working in the shallow waters of Takang,[49] but since 1970
official reports reveal that some of the veteran oil drillers have left
Takang for another new, unidentified oil field.[50] Thus the Chinese
oil workers resemble a special army corps that is transferred from
one field to another. The system, which is possible only for a totali-
tarian state, possesses the advantages of high mobility, well-
disciplined participants, and low training costs.

According to Western criteria, some Chinese developmental
policies may be viewed as irrational. The continuous effort to revive
the old deposits in Yumen and Yench'ang, which are both more than
35 years old, involves increasing costs and diminishing production.
Many of the old wells have reached the "economic limit"--the point
at which the unit costs are above the going prices. However, the
Chinese authorities persist in their policy of making maximum use of
resources by the process of secondary recovery, even when it goes
beyond the generally recognized limit of profitability. Hindsight
also suggests that the old Chinese enthusiasm for building thousands
of small shale-oil and coal-oil plants using primitive methods was
not justified by eventual profits.

NOTES

1. Yen Erh-wen, "Oil to Dominate Old China," China Reconstructs, April 1966, pp. 15-17.

2. Chao I-wen, Hsin-chung-kuo ti Kung-yeh (Peking: T'ung-chi Ch'u Pan-she, 1957), pp. 49-50.

3. China News Analysis (Hong Kong), no. 220 (March 14, 1958), p. 1.

4. The Science of Petroleum, vol. 1 (London, 1938), p. 139.

5. Ibid.

6. American Institute of Mining and Metallurgical Engineers, Transactions 68 (1922): 1109.

7. First Five-Year Plan for Development of the National Economy of the People's Republic of China, 1953-1957 (Peking: Foreign Languages Press, 1956), pp. 73-74.

8. A. A. Meyerhoff, "Development in Mainland China, 1949-1968," The American Association of Petroleum Geologists Bulletin 54, no. 8 (August 1970): 1567-80.

9. Shih-yu Kung-yeh T'ung-hsun, April 17, 1957, p. 1.

10. Shih-yu Kung-yeh T'ung hsun, no. 6, 1957, pp. 20-23.

11. Fu Chiao-chin, Shih-chieh Shih-yu Ti-li (Peking: Ko-hsueh Ch'u-pan-she, 1959).

12. Jen-min Shou-t'se 1958 [People's Handbook, 1958] (Peking: Ta-kung-pao she, 1958), p. 473.

13. Jen-min Jih-pao, May 1, 1959, p. 13.

14. Jen-min Jih-pao, February 14, 1960, p. 6.

15. "Oil Bases in the Desert," China Reconstructs 22, no. 1 (January 1973): 22.

16. Shih-yu Lien-chih, no. 3 (1958), p. 5.

17. Harold Lubell, Middle East Oil Crisis and Western Europe's Energy Supplies (Baltimore: The Johns Hopkins University Press, 1963), p. 123.

18. Shih-yu K'an-t'an, no. 18 (1958), pp. 7-9.

19. Shih-yu K'an-t'an, no. 4 (1960), pp. 12-13.

20. New China News Agency (Peking), June 27, 1960.

21. Kuang-ming Jih-pao, October 10, 1959; also Jen-mih Jih-pao, November 10, 1959, p. 2.

22. New China News Agency (Peking), July 28, 1959.

23. New China News Agency (Karamai), June 28, 1958.

24. Shih-yu K'an-t'an, no. 18 (1958), pp. 7-9.

25. Shih-yu Lien-chih, no. 5 (1958), p. 1, and no. 6 (1958), p. 1.

26. Shih-yu Lien-chih, no. 5 (1958), pp. 1-2.

27. Shih-yu Lien-chih, no. 5 (1957), p. 2.

28. Shih-yu K'an-t'an, no. 3 (1960), p. 1.

29. Kung-jen Jih-pao, March 5, 1958.

30. Shih-yu K'an-t'an, no. 4 (1958), p. 2.

31. Chung-kuo Hsin-wen, December 29, 1959.

32. Ta-kung Pao, December 21, 1959.

33. Meyerhoff, op. cit., pp. 1567-80.

34. Chung-kuo Hsin-wen, December 30, 1972, p. 12.

35. China Reconstructs, June 1971, p. 20; also Wilfred Burchett, "On the Tach'ing Oilfield," Eastern Horizon, no. 4 (1973), pp. 4-18.

36. Burchett, op. cit., pp. 4-18.

37. New China News Agency (Tach'ing), January 6, 1973.

38. Ho K'o-jen, "The Development in Red China's Petroleum Industry," Fei-ch'ing Yueh-pao (Taipei), March 1, 1968, pp. 52-65.

39. The Oil and Gas Journal (March 10, 1975), p. 42.

40. Peking Review, no. 1 (January 1974), p. 5.

41. Peking Review, no. 21 (May 24, 1974), pp. 15-17; also China Reconstructs, October 1974, p. 8.

42. Far Eastern Economic Review, May 14, 1973, p. 41.

43. K. P. Wang, "The Mineral Resource Base of Communist China," in An Economic Profile of Mainland China (Washington, D.C.: Government Printing Office, 1967), p. 186.

44. Peking Review, no. 1 (1972), p. 17.

45. New China News Agency (Tach'ing), January 2, 1973.

46. BBC, Daily Broadcast Report no. 4199 (1973), pp. A3-46.

47. M. A. Adelman, The World Petroleum Market (Baltimore: The Johns Hopkins University Press, 1972), p. 26.

48. China Pictorial (Peking), no. 2 (1974), p. 5.

49. The Wall Street Journal, August 15, 1974, p. 1.

50. China Reconstructs, October 1974, p. 8.

**THE GROWTH OF
CRUDE OIL OUTPUT**

THE DATA PROBLEM

Any quantitative study of the Chinese economy after 1949 immediately encounters the problem of obtaining and evaluating statistics. The data problem, which constitutes the major deterrent in
Chinese economic studies, has been dealt with in detail by several
economists elsewhere and need not be repeated here.[1] Primarily,
it involves three dimensions: availability, consistency, and reliability.

The availability of quantitative data for the Chinese economy
in general and for the petroleum industry in particular varies considerably from period to period. Data were comparatively abundant
between 1955 and 1960 but became basically absent during 1961-64,
the years following the failure of the Great Leap Forward. Statistics
in fragmentary form reappeared in 1965-66, when the economy recovered; disappeared once again during the turmoil years of the Cultural Revolution (1967-69); and reemerged following 1970, in most
instances in a selective and fragmentary form.

In the early years of Communist control (1949-54) little data
on petroleum were published, since the petroleum industry played
an insignificant role in the total Chinese fuel supply. The establishment of the Ministry of Petroleum Industry in 1955 reflected the
growing importance of the industry. In the following year the first
technical journal, the Shih-yu Kung-yeh T'ung-hsun (Bulletin of
Petroleum Industry), began semimonthly publication in Peking. It
contained major documents and technological information relevant
to the petroleum industry, including major conference reports, developmental plans, and news of the discovery and opening of new oil
fields.

In 1958 the bulletin was succeeded by two more specialized journals, Shih-yu k'an-t'an (Petroleum Prospecting), concentrating on geological prospecting and exploration, and Shih-yu Lien-chih (Petroleum Refinery), focusing on production and refining. Both, however, ceased publication in 1961.

These journals are invaluable for nonquantitative information, but little hard-core data pertaining to oil reserve estimates, crude oil output, refinery capacity, or petroleum products can be obtained from them. Whenever official statements relating to concrete statistics appeared, the figures were left completely blank in many places in the documents.

Nonetheless, in the yearly economic plans that were published regularly between 1954 and 1960, the annual crude oil output was officially reported. It was also summarized in Ten Great Years, a statistical document compiled by the State Statistical Bureau in 1959.[2]

Following the opening of Tach'ing in the early 1960s, the petroleum industry has been in the limelight. Numerous reports on Tach'ing have appeared in Chinese publications, most of them being folklore-type stories rather than concrete data. Even the location of Tach'ing was not revealed, not to mention its output and capacity, until July 1974, when a group of foreign correspondents was invited to visit it by the Peking authorities.

Between 1970 and 1974, two official figures on crude oil output were indirectly disclosed by Chinese Premier Chou En-lai when he granted interviews to foreign guests. In early 1971 Chou gave the figure of 20 million tons as the crude output for 1970, to the American journalist Edgar Snow.[3] In early 1974 Chou revealed the figure of 50 million tons as the crude output for 1973, to Japanese Foreign Minister Masayoshi Ohira.[4]

In June 1974 the Institute of International Relations in Taiwan published a secret document obtained from mainland China entitled "Outline of Education on Current Situation." This document had been compiled by The People's Liberation Army for the education of military officers and had been reprinted by the Kunming Military Region in April 1973. This secret document contained crude oil output figures for 1965 (11 million tons) and 1972 (45 million tons).[5] The output of crude oil between 1965 and 1974 can be derived from these two figures by applying percentage increases of crude oil as revealed in various official sources.

From cross-checks and backward and forward derivations, it appears that most of these official data are consistent. For instance, the 1965 and 1972 figures disclosed in the secret document correlate with the official report in Peking Review that "China's crude oil output in 1972 was four times the amount in 1965."[6] Calculated in a backward order, the 1972 output was officially stated as 16 percent over that in 1971,[7] which should be 38.8 million tons (45 million tons ÷ 116 = 38.8 million tons). This figure does not deviate much from another official statement, which gave 1971 crude oil output as increased more than 301 times over that in 1949.[8] Since the 1949 output was officially given as 121,000 tons, the 1971 output should be more than 36.5 million tons.

Again, when the 1972 figure of 45 million tons is used as the basis and calculated in a forward order, the 1973 output of 50 million tons as disclosed by Chou En-lai looks very reasonable, since the rate of growth was 11 percent, which is in accord with the 10 percent growth rate for Tach'ing in 1973.[9]

Chou's early revelation to Edgar Snow that the 1970 crude output was only 20 million tons is a noticeable inconsistency, incompatible with other official figures. For instance, according to Peking Review, the 1971 crude output increased by 28.6 percent over 1970.[10] Judging from the 1971 figures derived above (38.8 million tons, or 36.5 million tons), the 1970 output should be between 28.5 million tons and 30 million tons. This is much higher than Chou's figure.

Although the apparent contradiction between Chou's figure and other official figures warrants a cautious reservation about the reliability of the official statistics, by and large the aggregate oil output data contained in the secret document are in congruence with available information on the output of the major oil fields and can be taken as a basis for the derivation of other years. In an early study on the Chinese petroleum industry, K. C. Yeh also detected no apparent falsification in official petroleum output data, since adverse information, such as the failure to reach the First Five-Year Plan target, had been admitted, which would have been unnecessary if the data were falsified.[11] Yeh's observation is confirmed in recent findings. While the Chinese government tends to publish figures reflecting achievement, such as 1965 and 1972, it has generally withheld data for adverse years such as 1967 and 1968. This would suggest that a double-bookkeeping-type manipulation probably did not exist.

ESTIMATES OF CRUDE OIL OUTPUT

Official Statistics, 1949-60

Official statistics on crude-oil output during the 1949 to 1958 period were summarized in the Ten Great Years, as shown in Table 2.1.

The 1959-60 output was announced in early 1960 by Li Fu-chun, former chairman of the State Planning Committee, in his report on the 1960 national economic plan.[12] Since then there have been no continuous published official output statistics.

TABLE 2.1

Crude Oil Output, 1949-60

Year	Output (in tons)	Rate of Growth (in percent)
1949	121,000	--
1950	200,000	32
1951	305,000	52
1952	436,000	43
1953	622,000	42
1954	789,000	27
1955	966,000	22
1956	1,163,000	20
1957	1,458,000	25
1958	2,264,000	55
1959	3,700,000	63
1960	5,200,000	49

Source: People's Republic of China, State Statistical Bureau, Ten Great Years (Peking: Foreign Languages Press, 1960).

Independent Estimates, 1961-73

A wide gap in output data existed in the 1961-70 period. For 1961-64 and 1966-69 there were virtually no official figures concerning crude oil production. To fill the gap, many estimates have been made by specialists outside the Chinese mainland. Their estimates are summarized in Table 2.2, from which it is evident that the differences among the various estimates are quite substantial, from 35 percent to 80 percent in various years. Estimates made in Japan are invariably higher than those made in Taiwan. In the United States, K. P. Wang of the U.S. Bureau of Mines and Robert Michael Field of the Central Intelligence Agency made different estimates. In general, however, most of the independent estimates, when compared with the figures derived for 1965 and 1970 in the first section of this chapter, are on the lower side, as shown in Table 2.3.

In more recent years, as Japan and the Western countries have developed more contacts with China, estimates on Chinese crude oil output have come closer to official data. Table 2.4 summarizes these estimates.

TABLE 2.2

Estimates of Crude Oil Output in China, 1961-70, by Authors from the United States, France, the USSR, Taiwan, and Japan

(in millions of tons)

Year	United States		France	USSR	Taiwan		Japan		
	Wang	Field			Chang	Ho	Takagi	Tagawa	Kambara
1961	6.20	4.50	5.50	--	5.40	5.65	5.00	5.65	5.26
1962	6.80	5.00	5.80	5.80	6.00	6.25	6.25	6.25	5.83
1963	7.50	5.50	--	--	7.00	7.22	7.00	7.22	6.50
1964	8.50	6.90	--	--	8.50	8.90	8.40	8.90	6.90
1965	10.00	8.00	9.30	--	9.50	11.80	10.00	11.80	8.67
1966	13.00	10.00	11.00	12.00	11.00	13.00	12.00	13.00	12.37
1967	11.00	10.00	8.75	10.00	8.50	--	10.00	10.50	10.40
1968	15.00	11.00	12.00	--	9.50	--	12.00	13.60	12.40
1969	20.00	14.00	--	15.00	11.00	--	14.50	16.80	14.50
1970	--	18.00	20.00	18.00-19.00	14.50	--	18.20	20.00	20.00

Note: This table is representative, not exhaustive. There are many other estimates that are not included.

Sources: (1) for the United States, Yuan-li Wu, ed., China: A Handbook (New York: Praeger Publishers, 1973), p. 855; (2) for France, Michel Georges, "Petroleum Exports Determined by Political Goals," Joint Publication Research Service no. 61,933 (May 8, 1974): 1-10; (3) for the USSR, Far Eastern Economic Review (Hong Kong), January 29, 1972, p. 23; (4) for Taiwan, Chang Chun, "Peiping's Petroleum Industry: Growth and Future Development," Issues and Studies (Taipei) 10, no. 8 (May 1974): 49; and Ho Ko-jen, "Peiping's Petroleum Industry," Issues and Studies (Taipei) 4, no. 11 (August 1968): 30; (5) for Japan, Keizo Takagi, "Peiping's Oil Resources," (Tokyo) nos. 8-9 (1973), p. 79; Torn Tagawa, "China's Petroleum Industry," Chugoku-Keizai-Kenkyu-Geppo (Tokyo), July-September 1973, p. 5; Tatsu Kambara, "Petroleum Industry in China," Sekiyu-no-Kaihatsu (Tokyo), April, 1972, p. 21.

TABLE 2.3

Variations in Independent Estimates of
Crude Oil Output, 1961-70
(in millions of tons)

Year	High Estimate	Low Estimate	Mean
1961	6.20	4.50	5.40
1962	6.80	5.00	5.90
1963	7.50	5.50	6.70
1964	8.90	6.90	7.90
1965	11.80	8.00	9.90
1966	13.00	10.00	11.90
1967	11.00	8.50	9.90
1968	15.00	9.50	12.20
1969	20.00	11.00	15.10
1970	20.00	14.50	18.50

Source: Sources are given in Table 2.2.

In general, most of the independent estimates under review
did not give sources for their information; nor did they present the
methodology of their derivations, together with the rationale behind
the derivations. Since there is no consensus on the magnitude of the
output in various years, one can still make estimates based on various official sources.

Estimates for This Study, 1961-74

The Assumptions

My estimates of China's crude oil output during the 1961-74
period are based on the following assumptions:

1. Official data for the 1949-50 period are accepted as basically reliable, since domestic output plus imports came very close to
the estimated domestic consumption for this period.
2. The output figures for the years 1965 and 1972 as disclosed
by the secret document are also accepted as reliable, since they are
consistent with other official reports and since these data were intended for internal information rather than for outside propaganda.

TABLE 2.4

Estimates of Crude Oil Output, 1971-73

(in millions of tons)

Year	United States	France	USSR	Taiwan		Japan		
				Chang	Chien	Takagi	Koide	Ohno
1971	36.70	25.00	25.00	18.5	25.6	22.0	36.45	31.8
1972	43.00	--	29.00	21.5	29.7	25.0	42.30	36.9
1973	53.00	50.00	--	25.0	35.0	--	50.00 or more	--

Sources: (1) for the United States, U.S. Department of Commerce, Overseas Business Reports, June 1974, p. 20; (2) for France, Michel Georges, "Petroleum Exports Determined by Political Goals," Joint Publications Research Service 933, no. 61 (May 8, 1974): 1-10; (3) for the USSR, V. I. Akimov and M. M. Nikolsky, "Paking's Current Economic Situation," in Sekai Josey (Tokyo), April 5, 1973, p. 23, and Novasti, March 20, 1974; (4) for Taiwan, Chang Chun, "Peiping's Petroleum Industry: Growth and Future Development," Issues and Studies (Taipei) 10, no. 8 (May 1974): 49; and Chien Yuan-heng, "Chinese Communist Petroleum Industry as seen from the World-Wide Energy Crisis," Chung-kung yen-chin (Taipei), May 10, 1974, pp. 65-66; (5) for Japan, Keizo Takagi, Issues and Studies (Tokyo), nos. 8-9 (1973), p. 79; Yashio Koide, "China's Crude Oil Production," Pacific Community 5, no. 3 (April 1974): 469; Hideo Ohno, "Transition Period of China's Petroleum Industry," Chugoku Kogyo Tsushin, February 1973, p. 9.

3. The 1965 and 1972 figures are used as the central data from which the figures for other years are derived.

Crude Oil Output in 1960-64

The last official output figure for crude oil was the planned target of 5.2 million tons for 1960. A Russian source gave output for 1960 as 5.5 million tons.[13] Since this figure squares with the official announcement that the Second Five-Year Plan target of 5 to 6 million tons was met two years ahead of schedule,[14] the 5.5 million ton figure is adopted for 1960. In August 1960 the Soviet government suddenly withdrew the 1,390 Russian technicians who had been working in Chinese industries and suspended all technological and economic assistance. Because of the disruption of the Soviet aid, crude output during the 1961-62 period can not have increased markedly; but since the Tach'ing oil field produced .5 million tons of crude oil in 1960 and 1 million tons in 1961, the 1961 output should have at least increased by .5 million tons over 1960, to approximately 6 million tons.

The 1962 crude oil output was reported as 10.8 percent over 1961;[15] thus the 1962 output must have been 6.7 million tons, based on the estimate of 6 million tons for 1961. During 1959-61 China imported from the Soviet Union an average of 3 million tons of petroleum, including crude oil and oil products; by 1963 these imports were totally suspended. In the same year China imported 1 million tons of petroleum from Romania, which suggests that Chinese petroleum output must have reached 7.5 million tons in order to achieve basic self-sufficiency. When the Second People's Congress convened its 4th session on December 4, 1963, it was officially announced that the country had "basically achieved self-sufficiency in petroleum products." The operation of Tach'ing was identified as the decisive factor in rendering China self-sufficient in oil.[16] An estimate of 7.5 million tons of crude output for 1963 would appear to be reasonably sound and consistent with the limited information available.

In early 1975 an official source indicated that crude oil output in 1974 was 7.5 times that of 1964.[17] By this, the 1964 crude oil output can be derived as 8.5 million tons. This estimate is generally consistent with the 11 million tons figure revealed in the secret document for the following year.

Crude Oil Output in 1966-68

Another period when official data were extremely scarce was during the Cultural Revolution, from 1966 to 1968. In the early phase of the Cultural Revolution (May-November 1966), industrial centers were guarded from the Red Guard rampages, and the year-end official reports claimed a 20 percent rise in gross industrial output value, the sharpest rise in three years.[18] Crude oil production in 1966 was officially reported as having fulfilled the target 35 days ahead of schedule, and new capacity for crude oil and refining increased in that year, exceeding the total capacity increase during the Second Five-Year Plan.[19] This implies that crude oil output in 1966 may have increased at least 3 million tons. Since crude oil capacity in the Second Five-Year Plan increased by 4 million tons, it seems reasonable to assume that the 1966 output reached 14 million tons.

Early in 1967 a struggle for power broke out in almost every factory and mine. From May to September of that year the country's industrial centers, including Tach'ing and Yumen, were seriously disrupted. One official report shows that "in September 1967, the class enemies tried to disrupt production and created disorder in Tach'ing oil field."[20] Although it would be hard to assess the degree of disruptions precisely, scattered evidence shows that at least one quarter of the annual production was seriously affected. Crude oil output in 1967 may have declined as much as 20 to 25 percent.

Industrial output picked up rapidly in the second half of 1968 as the disturbances began to subside. The magnitude of the increase ranged from 50 to 100 percent compared with the first half of 1968, indicating the continuing sharp decline in the first half of the year and the rapid recovery in the second half.[21] Assuming a 40 percent rise for the entire year, the 1968 crude production would have approximated 15.4 million tons.

Crude Oil Output, 1969-74

Using the two benchmark figures of 1965 and 1972 as a basis, the crude output for the 1969-74 period can be derived as follows:

1. The 1969 output was officially reported as 88 percent higher than 1965, which should be approximately 20.68 million tons.[22]

2. The 1970 output was officially reported as 40.9 percent over 1969, which yields an output of 29.14 million tons.[23]

3. The 1971 output was reported as 28.6 percent over 1970, or 37.5 million tons.[24]

4. The 1972 output was given as 45 million tons in the secret document.[25]

5. The 1973 output was officially reported as 17 percent over that in 1972, or 52.7 million tons.[26]

6. The 1974 output was officially announced as 20 percent higher than 1973, or approximately 63.2 million tons.[27] Table 2.5 summarizes my estimates as compared with those of others.

TABLE 2.5

Comparison of the Author's Estimates on Crude Oil
Output with Other Estimates, 1961-74
(in millions of tons)

| Year | Author's Estimates | Other Estimates | | |
		High	Low	Mean
1961	6.00	6.20	4.50	5.40
1962	6.70	6.80	5.00	5.90
1963	7.50	7.50	5.50	6.70
1964	8.50	8.90	6.90	7.90
1965	11.00	11.80	8.00	9.90
1966	14.00	13.00	10.00	11.90
1967	11.00	11.00	8.50	9.90
1968	15.40	15.00	9.50	12.20
1969	20.70	20.00	11.00	15.10
1970	29.10	20.00	14.50	18.50
1971	37.50	36.70	18.50	26.16
1972	45.00	43.00	21.50	30.70
1973	53.00	53.00	25.00	41.00
1974	63.00	70.00	--	--

Sources: Author's estimates are made in the second section of
Chapter 2 of the text; other estimates, in addition to those calculated
by the author, are found in Tables 2.2 and 2.4.

Evaluation of the Estimates

If a comparison were made of my estimates of crude oil production for 1961-69 and those of others, mine would come closest to
the estimates of K. P. Wang. However, my rejection of Chou's
1970 figure and acceptance instead of the figures in the secret document result in a significant departure from most of the other estimates for oil produced in 1970-73.

First, based on my estimates, the increment of crude oil production between 1970 and 1973 was 70 percent, with an annual growth
rate of 19.4 percent. Other authors estimate the increment to be
121 percent, with an annual growth rate of 30.3 percent. These estimates, however, are incompatible with the growth rates of
Tach'ing, the country's most important oil-producing center during
this period. As shown in Table 2.6, Tach'ing output in the 1970-73

period increased 60 percent, with an average annual growth rate of 17 percent, which is very close to the national average.

Second, based on my estimate, the 1973 output increased 17 percent over 1972, which is in accord with official reports of an 11 percent growth rate for Tach'ing. [28] If the 20 million tons figure for 1970 is accepted, the derived 1972 output would be 29.8 million tons.

> 1971 output = 20 million tons x 128.6 percent
>   = 25.72 million tons
>
> 1972 output = 25.72 million tons x 116 percent
>   = 29.84 million tons

Given these figures, the 52.6 million tons figure for 1973 would represent a net increase of more than 20 million tons in a single year, a growth rate of 67 percent, which appears inconsistent with the patterns of the major oil fields.

The validity of my estimates of aggregate crude oil output can be further substantiated by examining estimates of the crude oil output of the major oil fields. Since the components should add up to the total, any significant discrepancy between these two sets of estimates would signal some flaws in the estimate.

In 1974, there were eight major natural oil fields and two synthetic oil bases in China, the former consisting of Tach'ing, Shengli, Takang, Yumen, Karamai, Tsaidam, Szechwan, and a new field called "70 oil field," the latter consisting of Fushun in Manchuria and Maoming in Kwangtung. Together these ten oil fields account for 95 percent of the total output. Of these oil fields, Tach'ing stands out as the most significant one.

The Tach'ing oil field in the Sungliao plain of Manchuria began its production in 1960. Between 1960 and 1971 the average annual growth of crude oil produced in Tach'ing reached 35.2 percent. [29] The 1971 output was reported to be more than four times that of 1963. [30] The growth rate slowed down in 1972 and 1973. In 1972 a new oil field was developed in the northern region of Tach'ing, resulting in a 64.6 percent increase of capacity in 1974. In the first six months of 1974, crude oil output in Tach'ing rose 24.7 percent over the corresponding period of 1973. [31] Based on scattered data, crude oil production at Tach'ing from its inception to 1974 can be estimated as shown in Table 2.6.

Although the annual output for the Tach'ing oil field is derived from various sources, the estimates are basically sustained by and dovetail with other evidence. One official statement indicates that the 1972 January-August crude oil output by the newly formed Number 5 extraction unit at Tach'ing was more than 17 times the nation's total output in 1949. [32] Since the 1949 national output was given as 121,000

tons,[33] this would amount to at least 2,057,000 tons. The output of
the Number 5 extraction unit for the year must have been about 3
million tons. Because there were five extraction units at Tach'ing
in 1972, total output for that year must have been more than 15 million tons, very close to my estimate of 15.3 million tons for that year.

TABLE 2.6

Estimated Crude Oil Output and Growth Rate of Output
at Tach'ing Oil Field, 1960-74

Year	Output (in millions of tons)	Growth Rate (in percent)	Index (1965 = 100)
1960	0.48	--	11
1961	1.08	125	25
1962	2.00	85	47
1963	2.66	33	63
1964	3.40	27	80
1965	4.25	25	100
1966	5.40	27	127
1967	5.80	7	136
1968	6.20	7	146
1969	8.50	37	200
1970	10.63	25	250
1971	13.29	25	313
1972	15.30	15	360
1973	17.00	11	400
1974	20.74	22	489

Sources: Figures are derived as described in Table 3.4, from
Chung-kuo Hsin-wen, December 30, 1972, p. 12; Chao Yu-sheng,
"The Tach'ing Oilfield," in Collected Documents of the First Sino-
American Conference on Mainland China (Taipei: Institute of Inter-
national Relations, 1971), pp. 795-820; Asahi Evening Press
(Tokyo), July 29, 1974; Peking Review, June 7, 1974, August 23,
1974, p. 16, and November 25, 1966; Chosa Geppo (Tokyo), March
1968; Ta-kung Pao (Hong Kong), January 19, 1972, p. 23; Jen-min
Jih-pao, August 23, 1973; New China News Agency, January 4,
1974; New York Times, December 16, 1974.

Moreover, according to another official source, the total crude oil produced at Tach'ing during the 1966-70 period amounted to 2.5 times that produced in 1960-65.[34] Since the derived total crude oil output for the 1960-65 period was 13.87 million tons and the derived aggregate output for the 1966-70 period was 36.53 million tons, the latter is 2.6 times the former, a ratio that comes very close to the 2.5 times as officially reported.

The output of the Tach'ing oil field is compared with the total output of China in Figure 2.1 and Table 2.7.

TABLE 2.7

Output of Tach'ing Crude Oil as Percentage of
Total Oil Production, 1960-74

Year	Total Output (in millions of tons)	Tach'ing Output (in millions of tons)	Tach'ing Output as Percent of Total
1960	5.50	0.48	8.7
1961	6.00	1.08	18.0
1962	6.70	2.00	29.0
1963	7.50	2.66	35.5
1964	8.50	3.40	40.0
1965	11.00	4.25	38.6
1966	14.00	5.40	38.6
1967	11.00	5.80	52.7
1968	15.40	6.20	40.3
1969	20.70	8.50	41.0
1970	29.10	10.63	36.5
1971	37.50	13.29	35.4
1972	45.00	15.30	34.0
1973	52.70	17.00	32.2
1974	63.20	20.80	33.0

Sources: Total crude output is estimated by the author as shown in Table 2.5; crude output from Tach'ing is estimated by the author as shown in Tables 2.6 and 3.4. Sources are given in Tables 2.2, 2.4, and 3.4.

FIGURE 2.1

Total Crude Oil Output and Output of Tach'ing Oil Field, 1961–74

The paucity of data prevents construction of a complete output series for the other oil fields following the procedures used for Tach'ing. It is possible, however, to interpolate from the limited data to make estimates for the output of major oil-producing centers during the 1965-74 period. Although the detailed derivations will be shown in Chapter 3, the findings are summarized in Table 2.8.

Table 2.8 reveals some discrepancy between the identified output of the major oil fields and the estimated total for various years, which is to be expected, since these two sets of estimates were derived from different sources. The highest discrepancy accounted for 7 percent of the estimated total. This may stem from some omissions in the oil field output, notably newly developed but still undisclosed oil fields. Recent reports reveal several oil fields under construction, including the Itu oil field in Shantung, the "57" oil field in Hupeh, the "70" oil field in Kwangtung, and several others. Since 93 percent of the total output is identifiable, the discrepancy would not appear to vitiate the overall findings.

THE PATTERN OF GROWTH

Based on the estimates presented in the preceding sections, it is possible to discern the patterns of growth in Chinese petroleum production during the past 25 years.

First, the growth of petroleum production has been very rapid compared with that of steel and other industrial output. (See Table 2.9.) In 1957 China produced 5.4 million tons of steel, but it delivered only 1.5 million tons of crude oil, a ratio of 3.6 to 1. In 1958 the most optimistic expectation was that the crude oil output would reach 6.5 million tons in 1962, 15 million tons in 1967, and 40 million tons in 1972.[35] This target was overfulfilled when the petroleum industry turned out an estimated 45 million tons of crude oil in 1972. During the same period, however, steel output grew from 5.4 million tons in 1957 to only 7 million in 1962, 10 million in 1967, and 23 million in 1972. In terms of tonnage, the 1972 crude oil output almost doubled the steel output. Taking the entire period from 1952 to 1974 into account, the 25.4 percent annual growth rate of crude oil is 70 percent higher than the 14.4 percent growth rate of steel and approximately three times the 8.4 percent growth rate of the total industrial output.

Second, the rate of growth of crude oil output is not uniform throughout the period. It was quite high during the rehabilitation period (1949-57) and rose to its highest growth rate during the Great Leap Forward (1958-60), but was sharply reduced in the wake of the Great Leap Forward, picking up rapidly in 1965 and 1966. The

TABLE 2.8

Estimated Crude Oil Production by Major Oil Fields, 1965-74
(in millions of tons)

Oil Fields	1965	1970	1971	1972	1973	1974
Northeast China						
Tach'ing	4.25	10.63	13.29	15.30	17.00	20.74
North China						
Takang	--	1.00	1.30	1.90	3.30	4.12
East China						
Shengli	0.60	3.50	5.50	7.20	8.80	10.21
Northwest China						
Yumen	1.50	1.80	2.00	2.30	2.50	2.65
Karamai	2.70	3.00	4.45	4.50	5.00	6.00
Tsidam	0.30	1.90	2.20	2.80	3.00	3.50
Yen Chang	0.15	0.18	0.18	0.20	0.40	0.40
Southwest China						
Szechwan	0.45	1.33	1.50	1.80	2.00	2.20
South China						
"70" Oil field	--	0.10	0.30	0.70	1.00	1.30
Other oil fields*	--	1.00	1.50	2.00	3.00	3.50
Shale oil	1.40	2.51	3.00	3.50	4.00	4.00
Statistical						
discrepancy	0.35	2.15	2.28	2.80	3.00	4.38
Estimated total						
crude output	11.00	29.10	37.50	45.00	53.00	63.00

*Other known oil fields include "913" field in Shantung, "57" field in Hupeh (Wen-hui pao [Hong Kong, July 31, 1974], p. 1), and Itu oil field in Shantung (Ming pao [Hong Kong, November 15, 1973]).

Sources: Tables 3.1 through 3.9.

growth rate between 1970 and 1974 was 20 percent higher than that between 1961 and 1969. Between 1952 and 1973 the growth rate of crude oil fluctuated in a five- or six-year cycle, as shown in Table 2.10 and Figure 2.2.

TABLE 2.9

Growth Rates of Crude Oil, Steel, and Industrial
Production in Various Years
(in percentage)

Period	Crude Oil	Crude Steel	Industrial Production
1949-60	42.0	63.0	30.0
1949-52	53.0	104.0	35.0
1953-57	27.4	32.1	18.0
1958-60	56.0	55.0	43.0
1961-69	16.8	8.7	4.7
1961-64	13.7	6.0	8.3
1965-69	17.1	6.8	1.0
1970-74	20.4	10.7	9.5*
1952-74	25.2	14.4	8.4
1957-74	24.5	10.1	6.5
1965-74	21.0	10.0	7.1

*1970-72.

Sources: (1) Figures for 1940-60 are from K. C. Yeh, Communist China's Petroleum Situation (Santa Monica, Calif.: Rand Corporation, 1962), p. 6; (2) crude oil figures for 1961-74 are from Table 2.7; (3) steel and industrial production figures are from U.S. Congress, Joint Economic Committee, People's Republic of China: An Economic Assessment, 92d Cong., 2d sess. (Washington, D.C.: Government Printing Office, 1972), p. 83; (4) figures for 1974 are estimated by the author on the basis that industrial output went up 10 percent in 1974.

TABLE 2.10

Annual Growth Rate of Crude Oil Output, 1950-74
(in percent)

Year	Growth Rate
1950	32.0
1951	52.0
1952	43.0
1953	42.0
1954	27.0
1955	22.0
1956	20.0
1957	25.0
1958	55.0
1959	63.0
1960	49.0
1961	9.0
1962	11.7
1963	12.0
1964	13.3
1965	29.4
1966	27.3
1967	-27.5
1968	40.0
1969	34.4
1970	40.6
1971	28.9
1972	20.0
1973	17.0
1974	20.0

Sources: These percentages are given in Table 2.1 and computed from the author's estimates of output shown in Table 2.5; sources are given in Table 2.1 and in the second section of the text of Chapter 2.

Underlying the cyclical fluctuations are two important factors, the discoveries of new major oil fields and the output fluctuations in the major oil fields. During the first cycle, 1950-56, since the emphasis was on exploration rather than on production, the growth rate of output declined steadily. The peak-year (1959) rate in the 1957-60 cycle coincided with the operation of the Karamai and Lenghu oil

FIGURE 2.2

Annual Growth Rates of Total Crude Oil Output
and Output of Tach'ing Oil Field, 1960-74

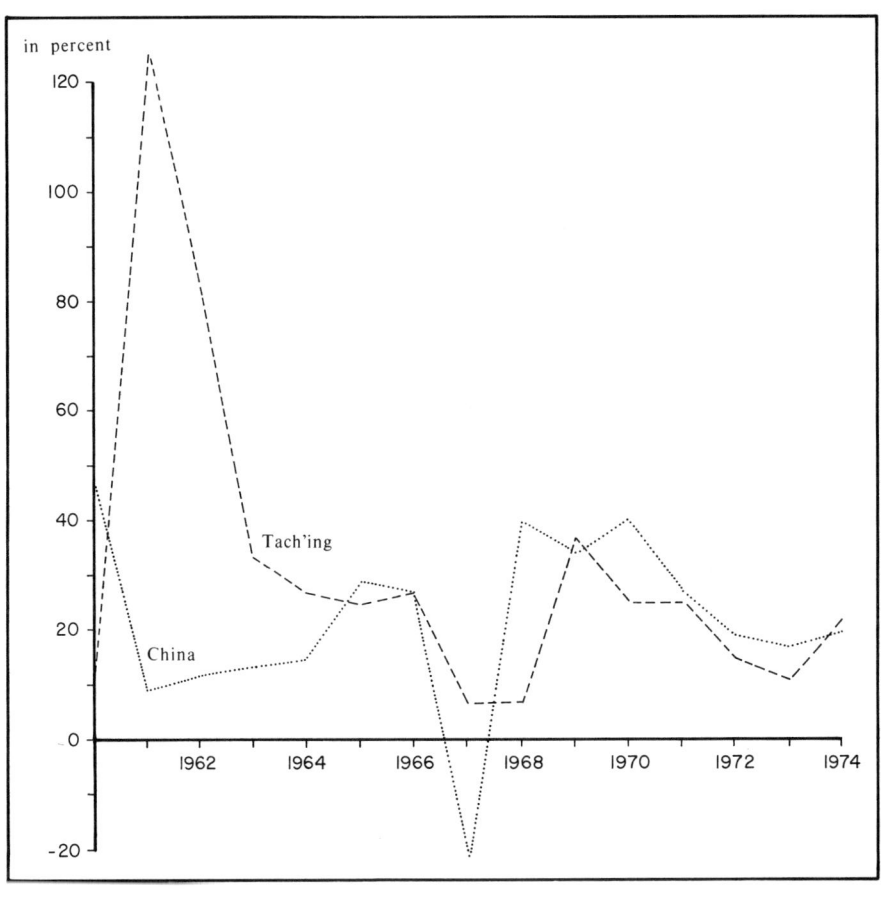

fields, both of which had been discovered in 1956 but started full-scale production in 1959. The peak year (1965) of the third cycle (1961–66) coincided with the full operation of Tach'ing, which contributed about 40 percent of the total output. The peak year (1970) of the fourth cycle (1967–73) coincided with production at the Shengli oil field. This clearly suggests that the most decisive factor of growth in petroleum output is the exploration activity of the preceding cycle, since the construction of a major oil field from the first test well to full production usually takes four to five years.

Within each cycle, the yearly variation is mainly affected by the output of the major oil fields. By comparing the annual growth rates of crude oil for the nation and for Tach'ing, a close relationship between these two can be discerned. (See Table 2.11.) However, the steady decline of the growth rate of Tach'ing output following 1969 has been more than offset by the growth in output of the newly developed fields at Shengli and Takang, keeping the overall growth rate of the petroleum industry above the growth rate of Tach'ing. The exploration of new oil fields in northern Tach'ing in 1972–73 brought the decline to a halt and once again began the upward swing of a new expansion, which may continue for several years.

TABLE 2.11

Annual Growth Rates of Total Crude Oil Output
and of Output of Tach'ing Oil Field, 1961–74

Year	Growth Rate of Total Output	Growth Rate of Tach'ing Output
1961	9.0	125.0
1962	11.7	85.0
1963	12.0	33.0
1964	13.3	27.0
1965	29.4	25.0
1966	27.3	27.0
1967	−21.5	7.0
1968	40.0	7.0
1969	34.4	37.0
1970	40.6	25.0
1971	28.9	25.0
1972	20.0	15.0
1973	17.0	11.0
1974	20.0	22.0

Sources: Figures are from Tables 2.5 and 2.6; sources are given in the second section of the text of Chapter 2 and in Table 2.6.

CRUDE OIL OUTPUT PROJECTIONS, 1975-85

In recent years, with the advent of the worldwide energy crisis, interest in China's potential as a crude oil exporter has increased. Projections of Chinese crude oil production and probable exports in the forthcoming decade have been made by various business groups, as well as by experts and observers in the United States and Japan. These "guesstimates" have frequently been revised with an upward bias that increasingly appears divorced from reality.

The figure game started with a spectacular projection made by Ryutaro Hasegawa, Chairman of the Japan-China Oil Import Council. Upon his return from a trip to China on April 15, 1974, Hasegawa staged an airport press conference in which he announced that China's annual crude oil output is expected to reach 400 million tons (8 million barrels per day) by 1980 and that one-quarter of this amount (2 million barrels per day) could be supplied to Japan.[36] Hasegawa proclaimed that these estimates were based on talks with Chinese authorities and his personal inspection of the Takang oil field. His statement became the prime source of other estimates.

In October 1974 Christopher Lewis, a British economist, writing in the Far Eastern Economic Review (Hong Kong), hypothesized that if the annual growth rate of 21 percent for the year 1973 were maintained, China's crude production would reach 65 million tons by 1974, exceed 78 million tons in 1975, and top the 200-million-ton mark by 1980. He further opined that China's crude output would equal 437 million tons by 1984.[37]

An optimistic estimate was also given by a Chinese-American scholar at the University of Chicago. Writing in the Los Angeles Times, Professor of History P. T. Ho waxed sanguine over China's petroleum prospects. Premising his study on "authoritative sources in Peking," Ho estimated that China's crude oil output would be 100 million tons in 1975 and 400 million tons by 1980. Moreover, he predicted that "by the early 1980's, China's oil exports might reach 300 million tons, or more than 2.1 billion barrels a year, an amount roughly equal to Iran's present annual production and more than Japan's present annual consumption."[38]

Ho's prediction was soon dwarfed by another spectacular projection made by Charles Abrams, chairman of the China Trade Corporation, who stated in late 1974 that "the oil reserves of the Chinese People's Republic exceeded those of the U.S. and the Middle East combined and by the 1990's, could supply all the world's oil need."[39] He even reportedly said that "China is currently supplying 70 million tons a year of oil to Japan and soon would be supplying 400 million tons."[40] Such a hyperbolic statement is inconsistent with the best information available on China's crude oil output for 1974; indeed, aggregate Chinese petroleum production from 1974 was only 63 million tons and would have been exceeded by these alleged exports to Japan by 7 million tons.

Since neither Hasegawa nor Abrams specified how they derived their figures, it is unwise to place much credence in them. However, Lewis's prediction, which is based on a 21 percent constant growth rate, deserves further comments.

As estimated in the preceding section, the average annual growth rate of crude oil in China during the 1952-74 period was 21 percent, which equals the growth rate projected by Lewis. However, the question is whether the growth rate for the past 22 years can be maintained through the forthcoming decade.

Historical records of the major oil-producing countries show that crude output has usually grown much more rapidly in the early phase of development and has subsequently been increased by relatively smaller growth rates. A survey of 13 major oil-producing countries, each with an annual output of more than 50 million tons in 1974, reveals, with the sole exception of Libya, that their long-term growth rates failed to exceed 15 percent. (See Table 2.12.)

TABLE 2.12

Long-Term Growth Rates of Crude Oil in
Six Major Oil-producing Countries

Country	Period	Average Annual Growth Rate (in percent)
Venezuela	1946-72	5.28
Kuwait	1955-72	5.95
USSR	1954-72	11.37
Saudi Arabia	1958-72	12.42
Iran	1960-72	13.01
Libya	1963-70	28.30

Source: Smil Vaclav, "Communist China's Oil Exports: A Critical Evaluation," Issues and Studies, March 1975, p. 75 notes.

The phenomenal growth rate of Libya was primarily the result of the simultaneous exploitation of her resources by a large number of multinational oil companies. During the 1960s in Libya, 37 oil companies commenced operations, among which were giants like Exxon, Mobil, Standard Oil of California, Texaco, and British

Petroleum.[41] For other large oil-producing nations, the typical
pace of oil industry expansion has been between 11 and 13 percent,
as illustrated by the USSR, Saudi Arabia, and Iran. Of these latter
three countries, the case of the USSR is particularly relevant to the
long-term projection of Chinese crude oil output, since neither the
Soviet Union nor China invited Western oil companies to help develop
their oil resources. In 1955 the Soviet Union produced 70.8 million
tons of crude oil, quantitatively similar to Chinese crude oil output
in 1975. Between 1955 and 1965 Soviet crude oil output rose from
70.8 million tons to 242.7 million tons, or 2.43 times, with an average annual growth rate of 13.1 percent. Moreover, the year-by-
year crude oil output evidenced a diminishing rate of growth instead
of a constant or increasing rate. (See Table 2.13.)

In the 1954-55 period, when the Soviet Union started expansion
of its crude oil output, its supporting industries were comparatively
much stronger than those of China in 1973-74. The Soviet Union
produced over 45 million tons of steel, 170 billion kilowatt hours of
electricity, and 731,000 tons of oil-production and refining equipment.[42] In 1973 China turned out only 25 million tons of steel, or
55 percent of the Soviet steel output of 1955. Chinese electricity
production in 1971 was estimated at 100 to 124.8 billion kilowatt
hours.[43] The supply of Chinese petroleum equipment was also considerably below the Soviet level. (See Chapters 4 and 5.) Given
these unfavorable considerations, the future growth rate of Chinese
crude oil production may be lower than the 13.1 percent rate previously achieved by the Soviet Union.

There are, however, two factors that would indicate a higher
growth rate in China's oil production, the first of which is the high
price of crude oil in current international markets, combined with
China's urgent need for foreign exchange to facilitate purchases of
Western machinery. The second factor is the development of new
technology and equipment by Western countries since the early 1960s,
which may also facilitate more rapid Chinese development. However,
constrained as China is by the limitations in her capital investment
and the scarcity of extraction, refining, and transportation facilities
(which will be discussed in subsequent chapters), a constant 21 percent annual growth rate for the forthcoming ten years seems optimistic indeed.

The experience of other major oil-producing countries tends
to indicate that as the production base is enlarged, the rate of growth
tends to decline. If the past patterns of other countries are indicative, one might more logically assume an annual growth rate of 20
percent for China's crude oil output for the 1975-77 period, 17 percent for 1978-80, 15 percent for 1981-83, and 12 percent after that.
This heroic assumption yields a forecast that places China's 1980

TABLE 2.13

Various Authors' Projections of Crude Oil Production in China, 1975-85, against Actual Output of USSR, 1955-65

| | Projections for China | | | | | | Actual Output of USSR | | |
| | According to Author | | From Source 1 | | From Source 2 | | | | |
Year	Production (in millions of metric tons)	Growth Rate (in percent)	Production (in millions of metric tons)	Growth Rate (in percent)	Production (in millions of metric tons)	Growth Rate (in percent)	Year	Production (in millions of metric tons)	Growth Rate (in percent)
1974	63.0	--	65.0	--	52.0	--	1955	70.8	20
1975	76.0	20	78.7	21	68.0	30	1956	83.8	18
1976	91.0	20	95.2	21	81.0	20	1957	98.3	17
1977	109.0	20	115.2	21	97.0	20	1958	113.2	15
1978	128.0	17	139.4	21	116.0	20	1959	129.6	15
1979	150.0	17	168.7	21	133.0	15	1960	147.8	14
1980	176.0	17	204.1	21	153.0	15	1961	166.1	12
1981	202.0	15	247.0	21	168.0	10	1962	186.2	12
1982	232.0	15	298.9	21	185.0	10	1963	206.1	11
1983	267.0	15	361.7	21			1964	223.6	9
1984	299.0	12	437.7	21			1965	242.9	9
1985	335.0	12	529.6	21					

Sources: Source 1 is Christopher Lewis, "Outlook Bright for Oil, Coal," Far Eastern Economic Review (Hong Kong), October 4, 1974, p. 22; source 2 is Tatsu Kambara, "The Petroleum Industry in China," The China Quarterly (London), December 1974, Table 8 on page 717; figures for the USSR are from Robert E. Ebel, Communist Trade in Oil and Gas (New York: Praeger Publishers, 1970), p. 40.

crude oil output at 176 million tons and her 1985 output at 335 million tons, with an average annual growth rate of 16 percent between 1975 and 1985. The 1985 projection of 335 million tons yields an annual growth rate for China's crude oil that is 20 percent higher than that of the Soviet Union during the 1955-65 decade, but 20 percent lower than the annual growth rate estimated by Lewis. A comparison of my projections with other estimates is summarized in Table 2.13.

NOTES

1. See Choh Ming Li, The Statistical Systems of Communist China (Berkeley: The University of California Press, 1964); Nai-ruenn Ch'en, Chinese Economic Statistics (Chicago: Aldine Publishing Company, 1968); Chu-yuan Cheng, Communist China's Economy 1949-1962 (South Orange, N.J.: Seton Hall University Press, 1963), Appendix; T. C. Liu and K. C. Yeh, The Economy of The Chinese Mainland (Princeton: Princeton University Press, 1965); and Alexander Eckstein, Communist China's Economic Growth and Foreign Trade (New York: McGraw-Hill, 1966), Appendix A.

2. People's Republic of China, State Statistical Bureau, Ten Great Years (Peking: Foreign Languages Press, 1960).

3. See Edgar Snow's article in The New Republic, March 27, 1971, p. 20.

4. New York Times, January 22, 1974.

5. The English-language version of this document appeared in Issues and Studies (Taipei), July 1974, p. 100.

6. "Rapid Expansion of Petroleum Industry," Peking Review, no. 39 (September 28, 1973), p. 22.

7. Ta-kung-pao (Hong Kong), December 27, 1972.

8. Peking Review, no. 39 (September 29, 1972), p. 12.

9. New China News Agency, January 4, 1974.

10. Peking Review, no. 47 (November 24, 1972).

11. K. C. Yeh, Communist China's Petroleum Situation (Santa Monica, Calif.: Rand Corporation, 1962), p. 1.

12. Peking Review, no. 14 (April 15, 1960), p. 13.

13. Far Eastern Economic Review (Hong Kong), January 10, 1963, p. 51.

14. China Reconstructs, April 1963, p. 6.

15. Chao Yun-ch'ing, "Communist China's Industry in 1963," Tsu-kuo [China Monthly] (Hong Kong), no. 2 (May 1, 1964), p. 11.

16. Jen-min Jih-pao, December 4, 1963; China Reconstructs, April 1967, p. 19.

17. Hsueh Shih-shih [Study of Current Events], no. 2 (1975); (Hong Kong: Wen Hui Pao), p. 36.

18. Chu-yuan Cheng, "The Effects of the Cultural Revolution on China's Machine-building Industry," Current Scene, January 1, 1970, pp. 3-5.

19. Wo-kuo Kuo-min Ching-chi Chuan-men Hsin-yueh-chin (Hong Kong: San-lien Su-tien, 1967), p. 22.

20. China Reconstructs, December 1968, p. 42.

21. Chu-yuan Cheng, "The Effects of the Cultural Revolution," op. cit., pp. 4-5.

22. The growth rate is given in an article by Hua Ching-yuan, "The Growth of Our Petroleum Industry," in Wo-men Cheng-tsai Ch'ien-chin (Peking: Jen-min Ch'u-pan-she, 1972), p. 46.

23. Ibid.

24. Ibid.

25. Issues and Studies, op. cit.

26. Hua Ching-yuan, "New Achievements in China's Oil Industry," China Foreign Trade, no. 1 (1975), p. 6.

27. Hsueh Shih-shih, no. 1 (1975), p. 1.

28. New China News Agency, December 31, 1973.

29. Chung-kuo Hsin-wen, December 30, 1972, p. 12.

30. Ibid.

31. Peking Review, no. 34 (August 23, 1974), p. 23.

32. Peking Review, September 29, 1972, p. 23.

33. Ten Great Years, op. cit., p. 95.

34. China Pictorial, no. 9 (1971), p. 5.

35. Shih-yu Lien-chih, no. 4 (1958), pp. 1-2.

36. BBC Summary of World Broadcasts, Far East, no. 4683 (August 21, 1974), p. C-1.

37. Christopher Lewis, "Outlook Bright for Oil, Coal," Far Eastern Economic Review (Hong Kong), October 4, 1974, p. 22.

38. Ping-ti Ho, "China's Resources Loom Large on World Stage: Huge Oil Deposits Could Weaken Arab Dominance," Los Angeles Times, October 13, 1974, pt. 6, p. 1.

39. Quoted by Richard C. Rowson, "China Tops Mideast, U.S. Oil Reserves," in the Overseas Press Club of America Bulletin 30, no. 1 (January 1, 1975): 1.

40. Ibid.

41. Smil Vaclav, "Communist China's Oil Exports: A Critical Evaluation," Issues and Studies (March 1975), p. 75.

42. Tsentralnoye Statisticheskoye Upravleniye SSSR, Narodnoe Khozyaystvo SSSR, 1922-1972 (Moskva, 1972); quoted from Smil Vaclav, op. cit., pp. 75-76.

43. The 100 billion kilowatt-hour figure is based on R. M. Field's article in U.S. Congress, Joint Economic Committee, People's Republic of China: An Economic Assessment, 92d Cong., 2d sess. (Washington, D.C.: Government Printing Office, 1972),

p. 83; the 124.8 billion figure is based on Thomas G. Rawski, "Measuring China's Industrial Performance, 1949-1973," paper presented at the Conference on Quantitative Measures of China's Economic Output, Washington, D.C., January 17-18, 1975, p. 48.

SOURCES OF INFORMATION

Publicizing of official data on China's major oil fields has varied erratically over time as well as in breadth of detail. The 1949-59 decade witnessed close collaboration between the Soviet Union and China with respect to exploration and development of China's infant petroleum industry. Russian geologists and petroleum experts participated in these oil explorations; indeed, the Sinkiang area was under the control of the Sino-Soviet Joint Stock Petroleum Company between 1950 and 1954. As a consequence, technical progress analyzing geological structure and sites of oil fields frequently appeared in both Soviet and Chinese journals.

Information with respect to oil resources in the northwest area has been publicized by the Chinese government to such an extent that a fairly detailed geographical account of the existing resources may be pieced together;* although it goes without saying that the frequent practice of Chinese officials of reporting that the annual output of a given oil field exceeds the previous six years' output of another field is frustrating to the analyst, but one must work with the information available.

---

*For instance, Y. I. Berizina's <u>Fuel and Power Base of the Chinese People's Republic</u> (Moscow, 1959), translated into <u>Joint Publications Research Service</u>, no. 3784, and Fu Chiao-chin, <u>Shih-chieh Shih-yu Ti-li</u> (Peking, K'o-hsueh, Ch'u-pan-she, 1959), provide detailed geological descriptions for the Dzungarian Basin, Tarim Basin, Tsaidam Basin, Chiu-ch'uan Basin, North Shensi Basin, and Szechwan Basin.

The Sino-Soviet split and the subsequent total withdrawal of Soviet geological technicians from China in August 1960 caused a radical reorientation in Chinese policy. The Soviet Union was no longer considered an ally but a potential aggressor, and the Chinese authorities became extremely sensitive about the security of the oil fields along the Sino-Soviet borders. The fear that the USSR might launch a military preemptive attack on the Chinese industrial bases has caused the Chinese authorities to clamp rigid censorship on information concerning the development of new oil fields in Manchuria and North China. The location of Tach'ing oil field was shrouded in secrecy for almost a decade; not until 1974 was the first international group of journalists permitted to visit it. The exact sites of Shengli and Takang are still not officially disclosed.

With the advent of the energy crisis, following the Arab embargo in October 1973, the Chinese leaders have adopted the strategy of using their oil potential for advancing their international position with regard to Japan and Southeast Asia. Publicity about new oil field development has begun to appear in Chinese journals and papers with increasing frequency, and foreign businessmen have been invited to tour some of the new oil fields and to inspect some of their facilities. This publicity has apparently been designed to promote foreign interest in Chinese oil resources; yet except for a continuous flow of heroic stories of the Tach'ing men, very little statistical data on production facilities, annual output, or oil resources has been revealed.

This lack of information makes the construction of a consistent output series for major oil fields extremely difficult. To bridge the numerous information gaps, many indirect derivations based on widely divergent sources have been utilized. As a consequence, my estimated sum of the crude output for individual fields in various years is, not surprisingly, different from national totals derived primarily from official statistics. The estimated output figures for individual fields are presented in subsequent sections of this chapter: when aggregated, they differ from national totals in two respects.

For 1949-65, the estimated annual crude oil output of individual fields, when aggregated, is larger than the national output figure released by Chinese officials. For the subsequent time-frame, 1966-74, the reverse is true: the aggregated individual oil field output figure is less than the officially proclaimed output total for the nation.

During the decade of Sino-Soviet cooperation, perhaps with the intention of repudiating the Western theory that China was an oil-impoverished nation, the Chinese government publicized new oil discoveries in hyperbolic fashion. Thus Karamai, in the

Dzungarian Basin, was described as the richest oil field in China,[1] and Tsaidam was called "China's Baku."[2] The official fanfare accorded the new oil fields and the highly inflated production targets accordingly assigned to them during the Great Leap Forward in 1958-59 led to gross overestimates of their output.

In sharp contrast, during 1966-74 Chinese authorities withheld information about the oil fields for national security considerations, and researchers outside China were left largely in the dark about newly developed oil fields in the coastal areas; this information curtain caused some underestimation of crude output for newly opened fields. For example, early estimates of crude oil production for Tach'ing and Shengli in 1965 were placed at 3,600,000 and 200,000 tons respectively; subsequent information indicated that Tach'ing produced 4,250,000 tons and Shengli had turned out some 680,000 tons for the year in question.[3] Moreover, production figures for some newly developed oil fields, such as the "913" oil field in Shantung, the "57" oil field in Hupei, and the "70" oil field in South China, were not released. Little wonder that national output estimates tended to exceed the output sum of the identified individual oil fields. (See Table 3.11.) Given these drawbacks and pitfalls, estimated output for each individual field may at best be taken as a rough approximation that indicates the direction of change but makes no claim to precise measurement.

MAJOR OIL FIELDS IN THE NORTHWEST REGION

Northwest China, which consists of the Sinkiang Uighur Autonomous Region and the three provinces of Kansu, Shensi, and Tsinghai, possesses the largest known oil reserve in China. The Chiu-ch'uan Basin of Kansu and the Dzungarian Basin of Sinkiang have long been recognized for their rich oil resources. The importance of the northwest region in China's petroleum industry was further strengthened in 1955 when the Tsaidam oil field in Tsinghai was explored. In 1959, before the opening of Tach'ing, the northwest oil fields were producing approximately 70 percent of China's crude oil, but the relative position of the northwest oil fields has been declining continuously since the opening of Tach'ing, Shengli, and Takang in the eastern coastal area. Thus, by 1974 only about 23 percent of China's crude oil output was coming from the northwest region. (See Table 3.11.)

The relative decline of the northwest oil fields primarily stems from three factors. First, the two older oil fields in the area, Yumen in Kansu and Yench'ang in Shensi, have been exploited since the Sino-Japanese war. As these reserves have dwindled, the

MAP 3.1

Major Oil Basins in Northwest China

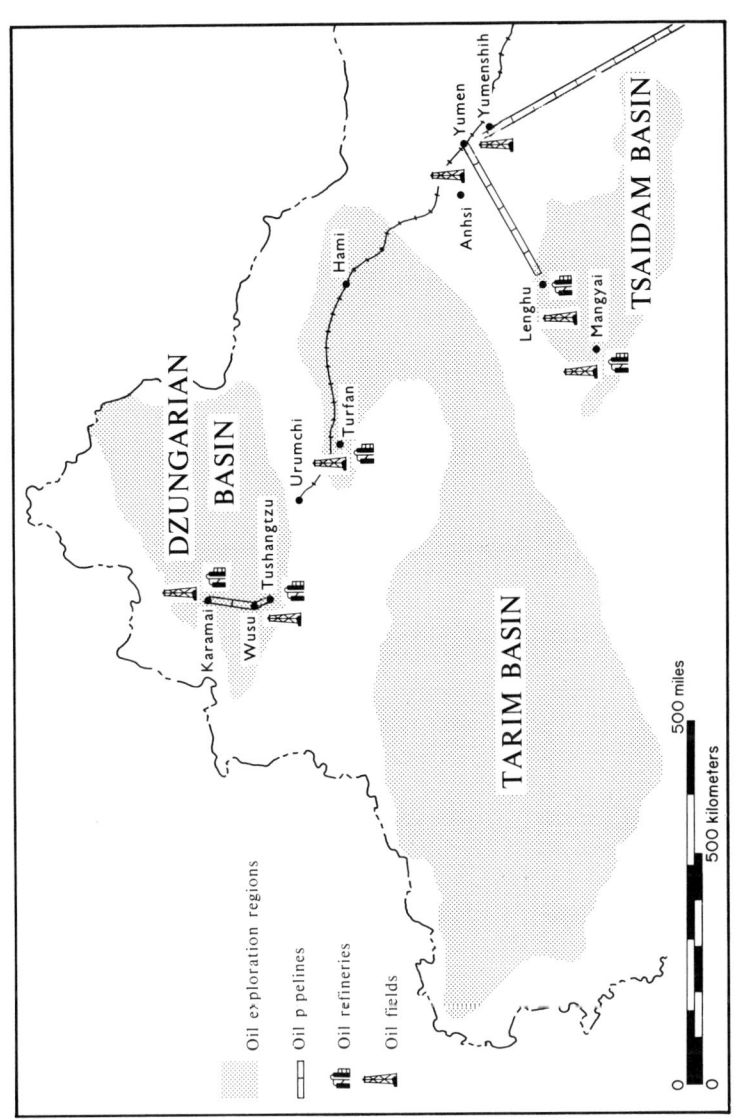

DZUNGARIAN
BASIN

Karamai
Wusu
Tushangtzu

Urumchi
Turfan

Hami

TARIM BASIN

Anhsi

Yumen
Yumenshih

Lenghu

Mangyai

TSAIDAM BASIN

Oil exploration regions
Oil p pelines
Oil refineries
Oil fields

500 miles

500 kilometers

TABLE 3.1

Estimated Crude Oil Output of Yumen Oil Fields, 1946-74

Year	Output (in tons)
1946	69,900[a]
1949	59,000[b]
1950	60,000[c]
1951	84,000[c]
1952	143,000[d]
1953	204,000[c]
1954	258,000[c]
1955	335,000[e]
1956	533,000[d]
1957	755,000[d]
1958	1,050,000[f]
1959	1,550,000[g]
1962	1,400,000[a]
1965	1,500,000[h]
1966	1,200,000[h]
1967	1,000,000[h]
1968	1,000,000[h]
1969	1,500,000[i]
1970	1,800,000[i]
1971	2,000,000[i]
1972	2,300,000[j]
1973	2,500,000[k]
1974	2,650,000[l]

[a]See Source 1.   [g]See Source 7.
[b]See Source 2.   [h]See Source 8.
[c]See Source 3.   [i]See Source 9.
[d]See Source 4.   [j]See Source 10.
[e]See Source 5.   [k]See Source 11.
[f]See Source 6.   [l]See Source 12.

Sources: (1) The figures for 1946 and 1962 are from Chang Kuei-sheng, Petroleum Resources and Production in Mainland China (Taipei: Institute of International Relations, 1963), pp. 9-10; (2) the figure for 1949 is derived from the official statement that the 1957 crude output of Yumen was ten times that in 1949; (3) the figures for 1950, 1951, 1953, and 1954 are derived from Ho Lan, Ti-i-ko Shih-yu-chi-ti-Yu-men (Shanghai: People's Arts Publishing House, 1956),

p. 1, according to which, crude output in Yumen, taking 1950 as 100, was 140 in 1951, 240 in 1952, 340 in 1953, and 430 in 1954;

(4) the figures for 1952, 1956, and 1957 were officially given in Kansu Provincial Statistical Bureau, "Communique on the Results of The Implementation of the National Economic Plan for 1957 in Kansu Province," Kansu Jih-pao (Kansu Daily) (Lanchow), June 12, 1958, p. 2; (5) the figure for 1955 is derived from Kuang-ming Jih-pao (Peking), October 1, 1956, which gave it as 30 percent over that of 1954;

(6) the figure for 1958 output was officially reported as 39 percent higher than that of 1957 in Chung-kuo Hsin-wen, February 26, 1959; (7) the 1958-59 output was reported as equivalent to five times the total crude produced in Yumen during the decade prior to 1949, which was officially given as 520,000 tons, making the 1958-59 output 2.6 million tons, and since the 1958 output was derived as 1,050,000 tons, output in 1959 must be 1,550,000 tons; (8) the figure for 1965 was given as 1.5 million tons in K. P. Wang, "The Mineral Resource Base of Communist China" in An Economic Profile of Mainland China, vol. 1 (Washington, D.C.: Government Printing Office, 1967), pp. 167-96, but in view of the large-scale transfer of equipment and technicians from Yumen to Tach'ing, output in the 1966-68 period must suffer some decline. It is assumed that the 1966 output dropped to 1.2 million tons and the 1967 and 1968 output further dropped to 1 million tons;

(9) According to Peking Review 5 (February 2, 1973): 26, output in 1971 was 32 percent higher than that for 1965; the 1971 output should be around 2 million tons. Output for 1971 was reportedly increased 11 percent over 1970 (Ching-chi Tao-pao, Hong Kong, January 19, 1972, p. 21). The 1970 output can be derived at 1.8 million tons. Based on the information of a 17 percent output increase for the first eight months in 1970 (Foreign Broadcast Information Service, Daily Report, October 20, 1970, p. H-1), the 1969 output can be derived at 1.5 million tons; (10) On the basis of a 14 percent output growth at the Shih-yu-kou oil field of Yumen in 1972 (BBC, Summary of World Broadcasts, Far East, W 697 A-8, November 1, 1972) and assuming that the growth rate was the same for the entire oil field, the 1972 output was estimated at 2.3 million tons.

(11) Based on the 9 percent increase of output for the first half year of 1973, output in 1973 was estimated at 2.5 million tons (Foreign Broadcast Information Service, Daily Report, August 17, 1973, p. H-1).

(12) Based on the information that the first seven months output in 1974 rose by 6.1 percent (Foreign Broadcast Information Service, Daily Report, September 11, 1974, p. M-1), the 1974 output was estimated at 2.65 million tons.

pressure from the subterranean oil has subsided, which tends to increase the cost of production.

Second, the two relatively new oil fields in this area, Karamai in Sinkiang and Tsaidam in Tsinghai, are located in sparsely settled areas where climates are extreme, water supplies short, and transportation facilities primitive. Under these conditions, exploration requires substantial overhead investment.

Third, since the opening of new oil fields near the east coast there has been a continuous outflow of equipment and technicians from the older northwest producing areas to support the new eastern oil fields and a consequent production decline for some years in the older area.

Since the eastern oil fields enjoy many comparative advantages in transportation and supporting infrastructure, the northwest oil fields will probably continue to play a declining role in the future of the Chinese petroleum industry.

The Yumen Oil Fields

The Yumen oil fields in the Chiu-ch'uan Basin of northern Kansu were the first large oil-producing center in China. The basin is bordered on the south by Chi-lien Mountain and on the north by Nei Shan. Several oil fields, known collectively as the Yumen oil fields, were found in the intermontane areas within the Chiuch'uan Basin. The most significant fields are at Lao-chun-miao, Ya-erh-hsia, and Pei-yang-ho. The reserves of the Yumen fields are estimated at nearly 540 million tons.

The Yumen oil fields were discovered in 1937, and active exploitation commenced in 1938. By 1946 production equaled 69,000 tons of oil, mostly from the Lao-chun-miao field. Between 1938 and 1947 more than 30 wells were drilled and an assortment of petroleum products were refined. Shortly after the Communist takeover, however, only six wells were left in operation. Restoration activities were started in 1949 with assistance from the USSR and the eastern European countries, and an expansion program was initiated in 1954. Basic construction investment in 1954 amounted to 2.7 times that of 1953, with 80 percent of it allocated for new construction projects. By 1957 producing wells numbered 170 and the crude oil output of the Yumen oil fields reportedly reached 755,000 tons, accounting for 89 percent of the natural crude oil produced in China and 52 percent of China's total petroleum output. (See Table 3.1.)

During the Great Leap Forward the crude output of Yumen rose precipitously. The 1958 output target was set at 39 percent over 1957, and total output in the 1958-59 period was reported as exceeding the ten years of output previous to 1949.[4] During 1958-59 intensive drilling led to the development of the Ya-erh-hsia section, which produced 30,000 tons of crude oil in 1959. This field extended from Yumen to Kiuchuan, the ancient city of Suchow. From Yumen, 548 miles of pipeline transports the crude to the main refinery at Lanchow.[5]

The withdrawal of the Soviet technicians in 1960 was a major cause of the decline in Yumen crude oil. Also, more than 30 years of operation had reduced the subterranean oil pressure. Lastly, Yumen's output was adversely affected by the withdrawal of large numbers of workers and equipment to support the new oil fields in Manchuria. Official reports reveal that between 1960 and 1973 Yumen exported 4.5 times as many workers to other oil fields as it employed in 1973.[6] Since there were 17,000 workers in Yumen in 1973, the workers that left Yumen to support other oil fields may be estimated at 76,500. At about the same time, Yumen transferred 75.5 percent of its equipment to the newly opened fields.[7]

Since 1969, by intensive use of secondary recovery methods, many abandoned oil wells have been brought back into production by water injection techniques. Official reports indicate that in 1970-73 the Yumen oil fields achieved a high and stable output for four consecutive years.[8] In 1974 an exploratory team, assigned to evaluate the oil potential of the area, disclosed that more than 2,400 promising oil wells and oil-bearing strata had been discovered.[9] However, crude oil output in the first seven months of 1974 rose by only 6.1 percent, a rate far lower than the 20 percent national average and indicative of Yumen's dwindling share of the national total.

The Karamai and Tushantzu Oil Fields

Prior to the opening of eastern oil fields, most Chinese and Soviet geologists believed that Sinkiang contained approximately 60 percent of China's total oil reserves.[10] Most of these reserves were found in two major oil fields located in the Dzungarian Basin: the Tushantzu oil field at the foot of the T'ienshan range and the Karamai oil field at the far northeast tip of Sinkiang.

The Tushantzu field was once believed by Chinese geologists to be the richest single oil field in China, having an estimated reserve of 840 million tons in 1951. The field was first explored in 1935, and by 1942 some 30 wells of differing depths had been drilled. Output remained very meager, with only 2,500 tons produced in that year. Although the field was expanded in 1950-52, its crude output was only 50,000 tons in 1952 and 94,500 in 1957. The Great Leap Forward output target for 1959 called for more than 200,000 tons, but adverse weather conditions and transportation bottlenecks apparently stymied achievement of this objective. Since that time Tushantzu has been noted for its two refineries, one of which was built with Soviet equipment in 1954 and enlarged in 1958 and 1960; it now commands an annual capacity of 1 million tons.[11]

The Karamai oil field stretches over a broad slope between the T'ienshan and the Altai-shan. Its operations commenced in October 1955, and 20 wells were each averaging 10-20 tons daily there by 1955.[12] The reserves at Karamai are reportedly much larger than the original estimate of 100 million tons.[13] The quality of the oil is superior to that of Yumen, having a low paraffin and sulphur content as well as a low freezing point. Commercial extraction began in 1958, and one of the wells recorded a daily flow exceeding 50 tons in April 1958--the same year the city of Karamai was established. The bulk of the oil produced in Karamai is shipped to Tushantzu for refining. A pipeline 147 kilometers (91.3 miles) long and capable of transporting 1.1 million tons annually was constructed in 1959 to connect Karamai with Tushantzu.

Oil-bearing strata were also discovered at Urho, 130 kilometers northeast of Karamai, and at Pai-chien-t'an, south of Urho. Output was targeted at 3 million tons for the entire Karamai complex in 1962. As a consequence of the Sino-Soviet split, the construction of the Lanchow-Sinkiang Railroad was interrupted; this in turn adversely affected the development of the Karamai-Urho-Tushantzu fields.

During 1959-62 Karamai was a major oil producer; 15 major construction projects were undertaken in this area in 1963. However, since the opening of the Tach'ing oil field the importance of the Karamai fields has continuously declined. Although the 1973 crude oil output in Karamai was officially reported as "substantially higher than before the Cultural Revolution,"[14] very little information is available about this oil field. A serious problem in the development of the Karamai-Urho fields has been inadequate water supply. Most of the water for oil-well injection comes from the Manass River by two water pipelines, but Manass dries up in the winter. Further expansion of the field would require new sources of water, which are not readily available in that arid climate. Currently, there are reports that oil workers in Karamai have unleashed a campaign to develop new fields.[15]

TABLE 3.2

Estimated Crude Oil Output of Karamai and
Tushantzu Oil Fields, 1951-74
(in thousands of tons)

Year	Karamai	Tushantzu
1951	--	7[a]
1952	--	50[b]
1956	5[b]	39[b]
1957	25[e]	69[c]
1958	250[e]	100[e]
1959	670[d]	NA
1960	800[e]	NA
1962	1,100[f]	NA
1963	1,500[g]	NA
1964	2,250[a]	NA
1965	2,700[h]	NA
1970	3,000[h]	NA
1971	4,450[h]	NA
1972	4,500[h]	NA
1973	5,000[i]	NA
1974	6,000[i]	NA

NA = Not available.

Sources: (a) The 1951 figure for Tushantzu and 1964 figure for
Karamai are from the Institute of Geological Survey, Academy of Industrial Technology, Collection of Latest Materials on China's Underground Resources (Tokyo, 1973), p. 307.

(b) The figures for Karamai in 1952 and Tushantzu in 1952 and
1956 are derived from the total natural crude output of the nation as
reported in official statistics by subtracting those produced at Yumen
and Yench'ang, the two other major oil fields operating at that time
(in tons) as follows:

	1949	1952	1956
Total Natural Output	70,000	195,000	589,000
Yumen	59,000	143,000	533,000
Yench'ang	1,000	2,000	13,000
Sinkiang	10,000	50,000	44,000

Totals are from Ch'en Nai-ruenn, Chinese Economic Statistics
(Chicago: Aldine, 1967), p. 187; Table 3.1; figures for Yench'ang
are from Shih-yu K'an-t'an, no. 3, 1959, p. 2; the Sinkiang figures
are the residuals. (continued)

(c) The 1957 and 1958 crude output of Sinkiang was given officially as 94,000 tons and 350,000 tons, in Ch'en, op. cit., p. 204.

(d) The 1959 planned output was 670,000 tons, in Survey of China Mainland Press (Hong Kong), No. 2132, November 6, 1959, p. 27.

(e) No official information for 1960-61. It is assumed that the field continued very moderate growth, as the manpower and equipment were concentrated in developing Tach'ing.

(f) Estimated on the basis of the figure for 1965 and the statement that output in 1965 was 2.6 times that in 1962 as reported in New China News Agency (Urumchi, Sinkiang), April 28, 1966, in Survey of China Mainland Press, No. 3689, April 28, 1966, p. 20.

(g) The 1963 output is given in K. P. Wang, "The Mineral Resource Base of Communist China," in An Economic Profile of Mainland China (Washington, D.C.: Government Printing Office, 1967), pp. 1, 294.

(h) The 1965, 1970, 1971, and 1972 figures are given on Ohno Hideo, "Transition Period of China's Petroleum Industry," Chugoko Koyyo Tsushin (Tokyo: February 1973), p. 9. The 1971 and 1972 figures are very close to estimates made by an expert in Taiwan who gave 3.8 million tons for 1971 and 4 million tons for 1972 (Chuan Wei, "A Critical Review on Recent Development of Petroleum Industry on Chinese Mainland," Ta-lu Ching-chi Yen-chiu 7, no. 1 [Taipei: January 1975]: 48). Both Ohno Hideo and Chuan Wei's estimates, however, are much higher than that estimated by Bobby A. Williams of the U.S. Central Intelligence Agency who put the 1971 and 1972 output at only 503,000 and 604,000 tons respectively. (See Williams, "The Chinese Petroleum Industry: Growth and Prospects" in Joint Economic Committee, Congress of the United States, China: A Reassessment of the Economy, July 1975, p. 259.) Since Williams' estimate for individual oil field output in 1974 is 19,564,000 tons less than his estimate for the national total, the Japanese estimate is accepted.

(i) The 1974 output is estimated on the basis of the statement that daily output at year-end 1974 was 200 percent higher than the 1965 (Foreign Broadcast Information Service Daily Report, December 17, 1974, p. M-5). The rate of output on an annual basis is believed to be less than the 200 percent increase. Assuming that the annual rate is 75 percent of the year-end rate, the 1974 output is estimated at 6 million tons. The 1974 output was reported as 20 percent higher than 1973 (Sinkiang Radio [Urumchi], December 12, 1973); the 1973 output should be around 5 million tons.

The Tsaidam Oil Field

The Tsaidam oil field, in the Tsinghai province of the Tibetan highlands, covers an area of 400 square kilometers. Chinese geologists evaluate Tsaidam as the most promising of China's oil fields.[16] Extensive prospecting began in 1954, and in the following year some 12,000 geological workers were dispatched to the western section of the basin; 118 promising oil formations had been found in the basin by 1958. The most significant oil-bearing area is located at Lenghu.

Prospecting work in Lenghu started in 1955. The high-quality crude oil produced in this area has a gasoline content of 40 percent. Both shallow and deep oils are present; some of the shallow oil traps are only 5 or 6 meters below the surface, while the deep ones lie approximately 2,885 to 2,905 meters down. Oil was first extracted in 1956 in an area known as the Number 4 Formation, which covers 100 square kilometers. In 1958 the Number 5 Formation was tapped, and since then it has been producing most of the oil in the Tsaidam Basin.[17] Reports indicate that in six months' time close to 100 wells were drilled, and peak daily production equaled 1,600 tons.[18]

Since operations in Lenghu began in 1959, crude oil output for the entire Tsaidam oil field increased eightfold.[19] However, the oil-bearing strata of Lenghu are elusively positioned and difficult to track, which accounts for the wide variability in its output. After several years of exploitation the pressure in the oil-bearing layers had dropped, and sand, collapsing into the well, had blocked the flow. In recent years some of the abandoned wells have been brought back into production by the injection of water. Within a year and a half, starting in March 1969, a new oil base in the western portion of the basin was opened. A recent report states that the confirmed oil-bearing area in the Tsaidam basin has been enlarged to four times its size between 1971 and 1973.[20] In 1972 output in Tsaidam resumed a sharp increase, partly because of the operation of the new oil structure[21] and partly because of a 40-kilometer pipeline providing water for oil fields completed in 1972.[22] In the past inadequate water supplies had caused a production bottleneck. (See Table 3.3.)

Since the mid-1960s a decade of construction has made Lenghu into one of the foremost oil producers of northwest China. It now boasts a gasoline refinery and 150 factories and mines that produce petroleum and petrochemical products and also cement, iron and steel, and machinery.[23]

TABLE 3.3

Estimated Crude Oil Output of Tsaidam Oil Field, 1958-74
(in thousands of tons)

Year	Output
1958	30[a]
1959	270[b]
1960	580[c]
1961	1,000[d]
1962	600[e]
1965	300[f]
1966	600[e]
1967	1,100[e]
1969	1,400[e]
1970	1,900[g]
1971	2,200[h]
1972	2,800[i]
1973	3,000[j]
1974	3,500[j]

Sources: (a) Figure for 1958 is from Wu Yuan-li, Economic Development and the Use of Energy Resources in Communist China (New York: Praeger, 1963), p. 182.

(b) Based on the information that crude output in Tsaidam increased eightfold after the operation of Lenghu oil field in early 1959 (Shih-yu K'an-t'an, no. 4 [1960], pp. 12-13).

(c) Figures for 1960 and 1972 are from Fuji Jahnalu, Chugoku Keizai no Genjo to Tembo, Tokyo, 1974.

(d) Figure for 1961 is from Ho K'o-jen, "The Development in Red China's Petroleum Industry," Fei-ch'ing Yueh-pao, March 1, 1968, pp. 52-65.

(e) Figures for 1962, 1966, 1967, and 1969 are from Institute of Geological Survey, Academy of Industrial Technology, Collection of Latest Materials on China's Underground Resources (Tokyo, 1973), p. 307.

(f) Figure for 1965 is from Ohno Hideo, "Transition Period of China's Petroleum Industry," Chugoku Kogyo Tsushin, February 1973, p. 9.

(g) 1970 figure is estimated based on the information that a new oil base was opened in 1970, in China Reconstructs, January 1973, p. 27.

(h) 1971 figure is from Jen-min Jih-pao, January 7, 1972, according to which oil extraction capacity went up 17 percent in 1971.

(i) Output in 1972 was officially reported as twice that of 1969 (Jen-min Jih-pao, March 12, 1973).

(j) Output in 1973 and 1974 is estimated based on recent report that confirmed oil-bearing area in the Tsaidam basin has been enlarged fourfold between 1971 and 1973 (Ta-kung pao, Hong Kong, July 7, 1974, p. 1, and Chung-kuo Hsin-Wen, March 11, 1973, p. 9).

MAJOR OIL FIELDS IN THE COASTAL AREAS

Since 1964 the focus of the Chinese petroleum industry has been shifted from northwestern China to the coastal areas. In 1960, when the Tach'ing oil field in Manchuria commenced operations, the coastal provinces accounted for only 10 percent of the nation's natural crude oil. The subsequent opening of Shengli and Takang in 1964 and 1965 greatly increased the relative importance of the coastal provinces. By 1974 these three east coast oil fields produced 63 percent of the nation's natural oil and almost all of its synthetic oil. Recent disclosures of oil-field development in Hupei, Shantung, and Kwangtung, if confirmed, will show a further advance in the relative share of coastal output. (See Maps 3.2 and 3.3, and Table 3.11.)

The Tach'ing Oil Field

Although the publicity limelight has been focused on Tach'ing's balanced integration of agricultural and industrial development since the early 1960s, the precise coordinates of this ideal type have been wrapped in secrecy. Foreigners were not allowed to visit the field until mid-1973. The first time an international group of journalists was invited to tour the field was in 1974. The reports of the journalists confirmed the initial assumption that Tach'ing was located between Harbin and Tsitsihar, two major heavy industry centers in Heilungkiang. The journalists further reported that this sprawling complex occupies a huge grassy plain about two and a half hours by rail northwest of Harbin. The oil field covers a distance of about 70 kilometers along the railway, from Langfeng to Taikang by way of Saertu and Changhulu in the northwest part of Heilungkiang. It consists of an eastern sector already developed and a northeastern sector currently under development.[24]

Based on the scattered evidence available, it would appear that the Tach'ing oil field is a kidney-shaped area lying between Fuyu, Taikang, and Machiayao. The field is a portion of the Sung-liao basin, which consists of an area of 250,000 square kilometers

MAP 3.2

Major Oil Basins in Northeast China

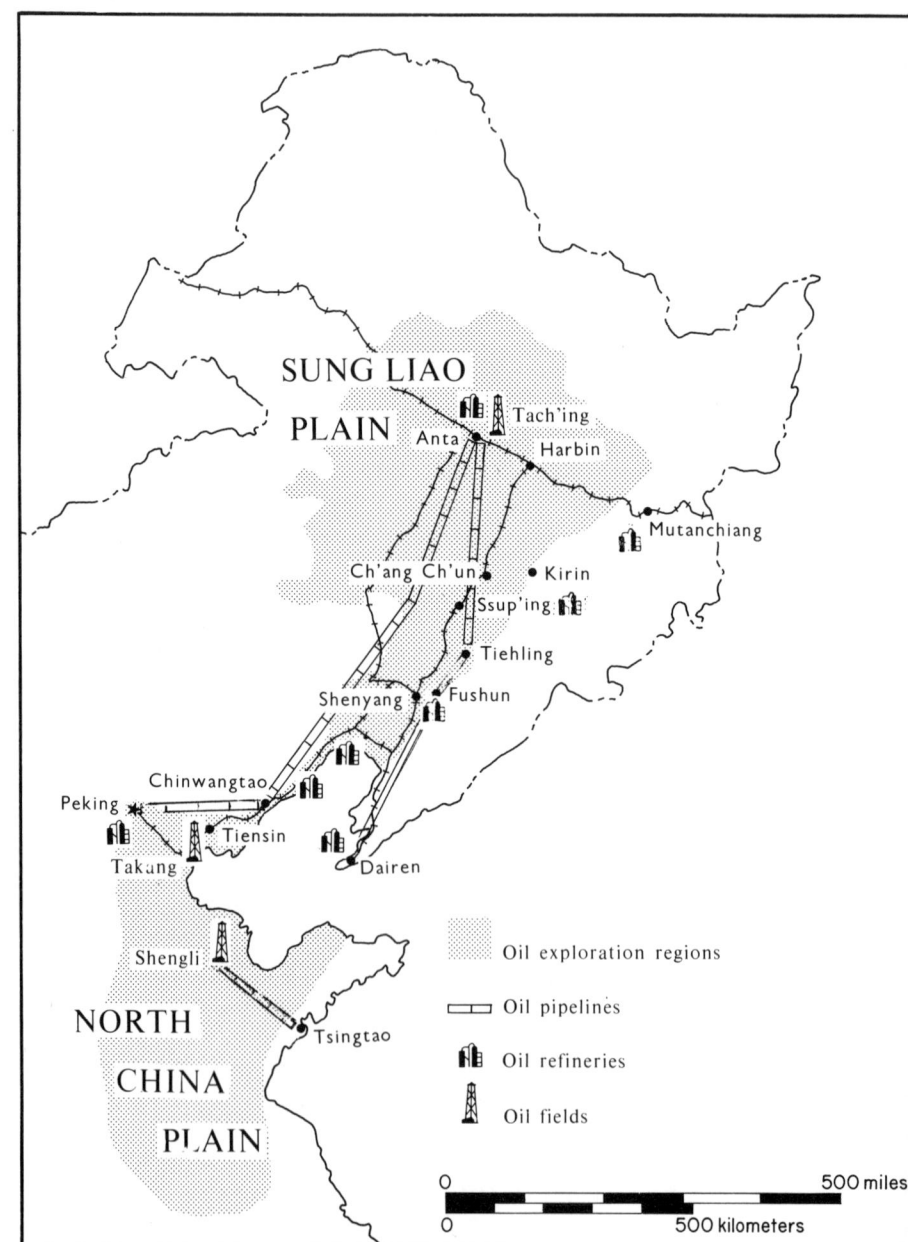

SUNG LIAO PLAIN

Tach'ing

Anta

Harbin

Mutanchiang

Ch'ang Ch'un

Kirin

Ssup'ing

Tiehling

Shenyang

Fushun

Chinwangtao

Peking

Tiensin

Takang

Dairen

Shengli

NORTH CHINA PLAIN

Tsingtao

Oil exploration regions

Oil pipelines

Oil refineries

Oil fields

0   500 miles

0   500 kilometers

(155,000 square miles). The most promising oil-bearing strata are
to be found in the depressed area near the center of the basin, cover-
ing more than 50,000 square kilometers.

According to a Hungarian expert who participated in the geophysical exploration work in Tach'ing from 1959 to 1962, the potential reserve of petroleum in Tach'ing was estimated at 1 billion tons.
By 1965 the proven petroleum reserves of the Tach'ing oil field may
already have exceeded 100 million tons.[25]

Because the oil field is accessible to the existing railroad network, the development pace has been very rapid. Between 1960 and
1971 the crude oil output in Tach'ing increased at an average annual
rate of 35.2 percent.[26] Between 1965 and 1973, crude oil output was
five times its 1965 output.[27] By 1965 Tach'ing's crude output had
overtaken Yumen, Karamai, Tsaidam, and the Central Szechwan oil
fields, to become the leading oil producer in China.

In April 1973 a new field in Tach'ing was opened that has an
annual production capability that exceeds the entire 1960-65 output
of Tach'ing. In June 1974 the daily output of crude oil in this new
field was officially reported as equivalent to the daily output of
Tach'ing for its first three years of operations.[28] (See Table 3.4.)

Construction of the Tach'ing refinery was started in April 1962,
and the first stage of construction was completed in October 1963.
In 1966 its refining capacity rose from 1 million to 2.5 million tons
a year, and by 1974 its capacity was rated at 5 million tons. In addition, a number of large- and medium-sized chemical facilities have
been added, with such products as synthetic ammonia, ammonium
nitrate, and acrylonitrile. Prior to 1975, more than half of Tach'ing's
crude oil output was transferred by single-track railway to refineries
at Fushun and Dairen. Another 2 million tons of crude was shipped
to the Peking Refinery.[29] The completion of the Tach'ing-Chinwangtao
pipeline in late 1973 facilitates crude oil exports to Japan, however.

Population in the Tach'ing oil field is reported to be 400,000,
of which 120,000 are oil field workers. In recent years, to facilitate
the development of other oil fields, Tach'ing has exported half of its
skilled work force as well as a large amount of its equipment to
Takang and Shengli. In this way, the Tach'ing facility not only en-
ables China to attain self-sufficiency in petroleum, but also provides
capital investment and know-how for the development of other new
fields. The profit accumulated by Tach'ing during its first 14 years
of operation is reportedly equivalent to 11 times the investment made
there by the Chinese government.[30]

TABLE 3.4

Estimated Crude Oil Output of Tach'ing Oil Field, 1960-74
(in thousands of tons)

Year	Output
1960	480
1961	1,080
1962	2,000
1963	2,660
1964	3,400
1965	4,250
1966	5,400
1967	5,800
1968	6,200
1969	8,500
1970	10,630
1971	13,290
1972	15,300
1973	17,000
1974	20,740

Sources and Notes: (1) The figure for 1960 is based on Chung-kuo Hsin-wen, December 30, 1972, p. 12, which said that between 1960 and 1971 the annual growth rate of crude oil at Tach'ing averaged 35.2 percent, while between 1965 and 1971 the crude output rose 212 percent. The 1965 output was about nine times that of 1960. Since the 1965 output was estimated at 4,250,000 tons, the 1960 output would approximate 480,000 tons.

(2) The 1961, 1962, and 1964 figures are estimated on the basis of number of oil wells in operation. According to Chao yu-sheng, "The Tach'ing Oilfield," in Collected Documents of the First Sino-American Conference on Mainland China (Taipei: Institute of International Relations, 1971), p. 804, there were 90 productive wells in Tach'ing in 1961, 200 in 1962, 320 in 1963, 450 in 1964, and 600 in 1965. Since 600 wells in 1965 produced 4.25 million tons of oil, one well on average yielded seven tons of oil. In 1963, 320 wells produced 2.66 million tons of oil, one well turning out about eight tons. This indicates that in the early years the wells were more productive. Using the 1963 figure as a basis, and assuming that output per well in 1961 was 12 tons and in 1962 was 10 tons, the 1961 output can be derived as 1.08 million tons and the 1962 output can be calculated as 2 million tons. Using 1965 figures as a basis, and assuming per-well output in 1964 at 7.5 tons, the 1964 output can be derived as 3.4 million tons.

(3) The figure for 1963 is based on the fact that output for 1971 was officially reported as having increased more than four times over 1963, in Chung-kuo Hsin-wen, December 30, 1972, p. 12. The 1963 output should be 2.66 million tons. (4) The 1973 output was given as 17 million tons in Asahi Evening Press (Tokyo), July 29, 1974, which was said to be four times the 1965 output in Peking Review, June 7, 1974, p. 16. The 1965 output should therefore be 4.25 million tons.

(5) The 1966 output was officially reported as 27 percent over 1965, which would be 5.4 million tons. Peking Review, November 25, 1966. (6) figures for 1967-69 are quoted from Akira Doi, "The Production and Price of Petroleum from Chinese Tach'ing Oilfield," Showa Dojin (Tokyo: August 1970), pp. 18-22. (7) The 1970 output was said to be 2.5 times that of 1965, or 10,630,000 tons, from Ta-kung-pao, September 12, 1971, p. 1.

(8) The 1971 output was increased by 25 percent over 1970, or 13,290,000 tons, from Ching-chi Tao-pao, January 19, 1972, p. 23. (9) Output in 1972 had risen 360 percent over that of 1965, or to 15.3 million tons. See Jen-min Jih-pao, August 23, 1973. (10) Output in 1973 had risen 10 percent over that of 1972. New China News Agency, January 4, 1974. (11) Peking Review, August 23, 1974, p. 23, reported that output in the first half of 1974 went up 24.7 percent over the corresponding period of 1973. The year-end figure, however, showed an increase of only 22 percent. See the New York Times, December 16, 1974.

The Shengli Oil Field

The Shengli oil field is the second largest oil field that was opened on the eastern coast during the 1960s. The field is located in Shantung Province near the mouth of Huang-ho (The Yellow River), encompassing several hsiens (counties) including Lichin, Chanhua, Yanghsin, Huimin, Linyi, Pohsing, and Kuangjao, and extending into the Paohai Bay. According to one official source, Shengli is situated in a section of the north China Basin. In the past, geologists held that no petroleum resources could possibly be found in northern China because of the absence of marine sediment of the mesozoic and cenozoic eras and the presence of igneous and metamorphic rocks over vast areas. However, inspired by the splendid success of Tach'ing, an extensive exploration was conducted in the early 1960s, resulting in the discovery of the Shengli oil field.[31]

Construction of Shengli began in 1964, and oil extraction started in the following year. At the first stage of development the operation was limited to the onshore area, but it was gradually extended into

Pohai Bay. Additional expansion led to the opening in 1971 of the area north of the Yellow River by some 10,000 workers exploring an expanse of weedy marshes fed by river water and tides.[32] The prospecting and development of two adjacent oil-producing areas began simultaneously in 1973.[33] Recent reports, based on Japanese sources, indicate that the oil-field expansion now includes the offshore area.

The scale of operations at Shengli is second only to that of Tach'ing. One official source claimed that in 1973 over 26,000 people were elected as advanced workers because of their work performance; it is estimated that in 1973 the labor force was approximately 100,000 people. Between 1965 and 1973 the crude oil output in Shengli was officially reported as having increased 13 times.[34] It was also revealed that a number of wells had been drilled and had been operating in the Pohai Bay area since 1966.[35]

A refinery has been completed at Hsintien along the Tsingtao-Tsinan Railway, and a pipeline network has been constructed to connect the oil field and the refinery. A portion of the crude is transferred by a new pipeline to the Peking General Petrochemical plant at Fenghuanling, 37 miles southwest of Peking. A recent report also indicated that a long-distance pipeline linking the oil field with the port of Huang-tao near Tsingtao, where new oil-handling facilities have been constructed, was completed and put into use in 1974.[36]

Although output at Shengli has grown rapidly in the past decade, production has not been stable. The presence of underground faults and complicated geological conditions has given rise to many new difficulties in prospecting and extracting the oil. In 1974, when the crude output of Tach'ing and Takang rose more than 20 percent, output in Shengli increased only 16 percent. (See Table 3.5.) However, during the first half of 1975, Shengli's new and expanded crude-production capacity rose 84.6 percent and crude production increased 44.3 percent. If the trend continues, Shengli may overtake Tach'ing in annual crude production in the early 1980s.[37]

Since the oil field is contiguous to major industrial centers near the east coast (Shanghai, Nanking, Tsinan, and Tsingtao in the south; Peking and Tientsin in the north), the further development of Shengli is strategically significant to China's future industrialization.

The Takang Oil Field

Takang, the third largest oil field developed in the 1960s, was first alluded to by officials in early 1974. The position and size of Takang oil field remains obscure. Some official sources indicate that it is situated 60 kilometers to the southeast of Tientsin,[38] while other official reports imply that it may cover most of the Gulf of Chihli-Huang-ho delta area, a basin of the offshore shelf.[39]

TABLE 3.5

Estimated Crude Oil Output of Shengli Oil Field, 1965-74
(in thousands of tons)

Year	Output
1965	600
1966	1,500
1967	NA
1968	NA
1969	2,000
1970	3,500
1971	5,500
1972	7,200
1973	8,800
1974	10,210

NA = Not available.

Sources: (1) The figure for 1965 is derived from China Reconstructs, 1974, p. 7, according to which output went up 13 times between 1965 and 1973. The 1965 figure is derived from the 1973 figure. (2) The figure for 1966 is from Chuan Wei, "A Critical Review on Recent Development of Petroleum Industry on Chinese Mainland," Ta-lu Ching-chi yen-chiu 7, no. 1 (January 1975): 44.

(3) The figure for 1969 is from Institute of Geological Survey, Academy of Industrial Technology, Collection of Latest Materials on China's Underground Resources (Tokyo, 1973), p. 307. (4) Figures for 1970-71 are from Ohno Hideo, "Transition Period of China's Petroleum Industry," Chugoku Kogyo Tsushin, February 1973, p. 9. (5) The crude oil output in Shantung Province was reported as increased by over 30 percent in 1972. Since the major oil field in Shantung was Shengli, it is assumed that the figure indicates the growth rate of Shengli oil field.

(6) According to New China News Agency (Peking), February 26, 1974, the major drilling team of the Shengli oil field set a national drilling record of 150,105 meters in 1973. This was more than twice the aggregate drilling footage of the 42 years before 1949. If crude oil output is assumed to correspond to drilling footage, the 1973 crude oil output at the Shengli oil field should approach 9 million tons, since total crude oil produced in the pre-1949 era was officially reported as 2.95 million tons, in Chung-kuo Hsin-wen, September 1, 1966.

(7) Output in 1974 was 16 percent higher than in 1973, according to Hsueh Shih-shih (Hong Kong), no. 1 (1975), p. 3. This estimate is 20 percent lower than some Japanese estimates. For instance, Yoshio Koide, a Japanese government officer, suggested that the crude output of Shengli in 1974 was probably around 12 million tons, in "China's Crude Oil Production," Pacific Community 5, no. 3 (April 1974): 466. Since this article was published in early 1974, this figure can only be taken as speculative. Experts in Taiwan, however, gave a much lower estimate, of about 5 million tons, for 1974. See Chuan Wei, op. cit., p. 44. In view of the number of wells (more than 2,000) and the number of workers (more than 100,000), an estimated output of around 10 million tons should not be too far from reality. More recently a Japanese source reported that the output of Shengli in 1974 was more than 10 million tons. See The Oil and Gas Journal, March 10, 1975, p. 42.

The field was prospected and commenced operations in 1964, paralleling the development of Shengli. In the spring of 1964 nearly 10,000 oil workers in Tach'ing were dispatched southward to the barren salt flats near Tientsin in order to explore the field. After the first well at Takang gushed oil in 1964, a series of deep exploratory wells were sunk to determine the area of the oil-bearing structure. It was found that the entire field was broken by faults. Some compared the area to an underground plate smashed into pieces, with so many faults that no regular pattern could be discerned.[40]

It would appear that the unique geological structure of Takang and China's lack of experience with offshore drilling prevented a rapid growth of crude oil output in this area. Unlike Tach'ing and Shengli, which achieved early success, the Takang has been under rather protracted development. Between 1967 and 1973, although its production of crude oil increased at an average rate of 60.9 percent a year, its total output over all eight years was only 3.1 times the total for the whole of China between 1907 to 1949.[41] This would imply that the aggregate crude output in Takang in the 1966-73 period amounted to approximately 9 million tons, an amount equal to the output of Shengli in 1973 alone. (See Table 3.6.)

Nevertheless, the long-future of the Takang oil field appears quite promising. According to one official report, explorations have verified that Takang abounds in oil and gas. The report further states that Takang possesses thick oil-bearing rock layers with good permeability.[42] One Japanese source claimed that the thickness of the oil-bearing layers averaged 2,000 meters (6,600 feet). The quality of Takang oil is superior to that of Tach'ing, having a sulphur content of .08 percent, as compared with the average of 2.5 percent for Arabian oil.[43] Moreover, where Takang extends into the offshore area, its output potential is believed to be far greater than that of Tach'ing.

The development of Takang is limited by the availability of offshore drilling equipment and know-how, as well as by inadequate transportation facilities. Since 1969 China has solicited foreign equipment and technology; in that year she placed an order to the Ishikawajima-Harima Heavy Industries of Japan for a sea-bed drilling ship. In September 1972 Japan delivered to China the Fuji, a drilling ship that cost $8.7 million (2,600 million yen). Information from Japan and the West indicated that drilling in the Pohai Gulf has been intensified since 1974. By mid-1975 China possessed six offshore rigs. The existing mobile fleet includes three Japanese-made rigs--the Fuji and two other second-hand rigs bought from Mitsubishi in 1973, one a jack-up and the other, a semisubmersible. China has three locally built mobile rigs named Pin-hai 1, Pin-hai 2, and Pin-hai 3, each capable of drilling to 2,200 meters in 30-meter-deep

waters. Further evidence of China's offshore push lies in its order
for supply boats: eight vessels from Denmark and five from Japan.[44]
In August 1974 Western sources reported that these new fields located in shallow water employed 40,000 workers.[45] These bits of
evidence suggest that the development of the Takang oil fields may
be accelerated in the coming years.

TABLE 3.6

Estimated Crude Oil Output of Takang Oil Field, 1966-74
(in thousands of tons)

Year	Output
1966	100
1967	200
1968	400
1969	500
1970	1,000
1971	1,300
1972	1,900
1973	3,300
1974	4,120

Sources: (1) The figures for 1966-69, 1971, and 1973 are derived from China Reconstructs, October 1974, p. 8, according to
which the crude oil output of Takang between 1966 and 1973 was 3.1
times the total for the whole of China in the 42 years from 1907 to
1949. This would be 8,680,000 tons. The same source indicates
that the average annual growth rate between 1967 and 1973 was 60.9
percent. This means that the 1973 output was 1,677 percent of 1967.
Estimated figures for 1968, 1969, and 1971 are based on circumstan-
tial evidence. However, the total estimated output for 1966-72 added
up to 8.7 million tons. The estimated output for 1973 equals 1,650
percent of the 1967 output and is consistent with the 1,677 percent
derived from the report in China Reconstructs.
　　(2) In 1970 the crude oil output was reportedly twice what it
had been in 1969. New China News Agency (Tientsin), May 28, 1974.
(3) According to Chung-kuo Hsin-wen, June 14, 1973, p. 1, from
1966, when Takang went into operation, until 1972, the crude oil output in the Tientsin area increased 18 times. The 1972 output appears
to be around 1.9 million tons. (4) In 1974 the crude oil output of
Takang was 24.7 percent over that of 1973, according to Hsueh Shih-
shih, no. 1 (1975), p. 7.

The Szechwan Oil Field

The Szechwan oil field is located in the middle portion of the Szechwan Basin in Southwest China. Surrounded by high mountains, the area has widespread oil, gas, and bituminous resources. During the Sino-Japanese war, 1937-45, preliminary drilling was done in many parts of the basin. Large-scale exploration was started in late 1957 and early 1958 with technological aid from the Soviet Union. Large oil reserves in three formations, Penlaichen, Nanch'ung, and Lungnussu, in the central Szechwan platform, were discovered as a result of the exploration. Deep within the platform there exists several thousand meters of shallow marine-type Paleozoic and marine-origin Permian formations. Cretaceous and Jurassic sediments are thick, totaling some 3,000 meters.

The Lungnussu anticline is located in the eastern part of the platform, about 150 kilometers north of Ch'ungking, the wartime capital of China. The Chialing River runs throughout the Lungnussu formation, with the two hsiens of Wusheng and Yuehch'ih in Szechwan on either side.[46] Recoverable reserves in this formation were estimated at 606 million barrels (83 million tons).[47]

Not far from and to the northeast of the Lungnussu formation is the Nanch'ung formation, which is located in the hsiens of Nanch'ung and Penglai. The center of the Nanch'ung formation is situated in the Tungkuan chen (township) of Nanch'ung Hsien. The formation covers a wide region totaling 800 square kilometers. West of the Lungnussu anticline is the Penglaichen formation. The boundary of the formation encloses about 400 square kilometers.

The triangular area formed by the three structures covers an area of several thousand square kilometers. Oil ultimately recoverable from these fields is estimated at more than a billion barrels.[48]

In 1958 there were 22 oil wells and 14 natural-gas wells in the Szechwan oil field, producing 120,000 tons of crude and more than 700 million cubic meters of natural gas. The crude oil is of very good quality, and the specific gravity is less than .86. This extensive area of rich deposits made the field a most promising enterprise; however, the production depth exceeds 1,500 meters, and consequently the well pressure runs very high, necessitating the use of heavy clay to constrain the excessive pressure. Frequent drilling accidents were accompanied by escalating production costs, with the consequence that the development of the Szechwan oil field lagged behind that of the other major fields. As of 1966 only 75 oil wells were in operation, with an aggregate annual output of 930,000 tons.[49] The current emphasis on development of oil fields near the eastern coast appears to have preempted much of the attention and publicity that might otherwise have been focused on the Szechwan oil field; thus current data on Szechwan are obscure, and when the Szechwan provincial government published its achievements under the 1971 industrial plan, crude oil production was not mentioned.[50] From this report one might be tempted to conclude that little progress was made in Szechwan oil production, but current output guesstimates still range from 2 to 2.2 million tons annual output. (See Table 3.7.)

TABLE 3.7

Estimated Crude Oil Output of Szechwan Oil Field, 1958-74
(in thousands of tons)

Year	Output
1958	120
1959	250
1960	340
1961	440
1962	NA
1963	NA
1964	730
1965	450
1966	930
1967	1,330
1968	NA
1969	1,000
1970	1,330
1971	1,500
1972	1,800
1973	2,000
1974	2,200

NA = Not available.

Sources: (1) Ho Ko-jen, "Peiping's Petroleum Industry," Issues and Studies 4, no. 11 (August 1968): 28-29. (2) Ohno Hideo, "Transition Period of China's Petroleum Industry," Chugoku Kogyo Tsushin, February 1973, p. 9; (3) Institute of Geological Survey, Academy of Industrial Technology, Collection of Latest Materials on China's Underground Resources (Tokyo, 1973), p. 307; (4) Tatsu Kambara, "Petroleum Industry in China," Sekiyu-no-Kaihatsu, April 1972, p. 27; (5) Chien Yuan-heng, "Chinese Communist Petroleum Industry as Seen from the Worldwide Energy Crisis," Chung-kung Yen-chiu, May 10, 1974, p. 63 which placed recent output at around 2 million tons.

In 1959 a refinery with a processing capacity of 900,000 tons was reportedly constructed at Nanch'ung. The balance of crude output from Szechwan was brought to the refineries in Shanghai by oil tankers on the Yangtze River.

THE SHALE OIL REFINERIES

China possesses large deposits of oil shale distributed over much of the country. According to a geological survey of the prewar era, oil shale reserves exist in the provinces of Heilungkiang, Kirin, Liaoning, Szechwan, Shensi, Kwangtung, and Hunan. Total oil shale reserves were estimated at 11,892 million tons, capable of yielding 521 million tons of shale oil.[51]

In 1958 the Chinese government published a new estimate of 360,000 million tons of shale oil reserves for the 21 provinces, a tremendous increase over the 1949 estimate.[52]

In 1966, according to a Taiwan source, the oil shale reserves were updated to 380,000 million tons.[53] The geographical distribution of the reserves is shown in Table 3.8.

Despite their widespread existence, most of the oil shale deposits remain unexploited. The largest shale oil extraction facilities were built at Fushun, in Liaoning province. The Number 1 and Number 2 refineries at Fushun built by the Japanese during the 1930s were partly stripped by the Russians in 1945. Restoration began in 1950, and by 1955 the Number 1 plant possessed a refining capacity of 675,000 tons and the Number 2 plant, 202,000 tons. In addition, new facilities were built at Huatien in Kirin, and a major refinery south of the Yangtze River was constructed at Maoming in Kwangtung Province in 1958.

Before 1958 shale oil accounted for almost half of China's total crude production; since that time the relative share has steadily declined as natural oil fields have been developed. By 1974 shale oil contributed less than 10 percent of China's total crude oil output.

The Maoming Shale Oil Plant

The Maoming Shale-Oil Company, located on the banks of the Chien-chiang on the coastline of Kwangtung province in southern China, is currently the largest single shale-oil producing center in China. Underground deposits of oil shale in the three mining areas cover 400 square kilometers with a probable reserve of 6.2 billion tons.

TABLE 3.8

Estimated Oil Shale Reserves in China, 1966

Location	Oil Content (in percent)	Oil Shale Deposits (in billions of tons)
Northeast		
Fushun (Liaoning)	6-10	5.60
Huatien (Kirin)	8-20	0.54
Peian (Heilungkiang)	6-8	130.00
Harbin (Heilungkiang)	4-8	2.00
Mutankiang (Heilungkiang)	10-20	0.18
North		
Ch'engte (Hopei)	5-11	100.0
Shansi	10-20	2.00
Inner Mongolia	6-8	1.00
Central		
Tungpo (Honan)	8-15	0.25
South		
Maoming (Kwangtung)	7-8	6.25
Kwangsi	10-27	0.30
Northwest		
Shensi	4-5	6.40
Urumchi (Sinkiang)	6-10	0.21
Southwest		
Yunnan	25	0.10
Other areas (unidentified)		135.17
Total		380.00

Source: China Petroleum Corporation, Ta-lu Shih-yu Kung-yeh Hsian-shih [The Present Status of the Petroleum Industry in Mainland China] (Taipei: the Corporation, 1968), pp. 61-63.

Construction of the plant was initiated in 1958. In the original plan it was contemplated that by the time the first stage of construction of the ancillary overhead facilities was completed, the annual crude oil output would have reached 1 million tons. However, the plan was shortly superseded by the Great Leap Forward, which increased the output target for 1960 to 2 million tons.[54] The ambitious goals of the revised plan failed to materialize, however, since the program soon collapsed.

During the first phase of construction much of the investment was concentrated in overhead capital in the form of the Lienchiang-Maoming branch of the Lichan Railway and the Maoming-Sansui branch railway. The Maoming Shale-Oil Company covered an area of more than two square kilometers and was comprised of hundreds of antiquated dry–distillation ovens for shale oil extraction, each of which could process 200-300 tons of shale daily, yielding some ten tons of crude oil. In addition, processing installations for extracting light and heavy diesel oil, gasoline, and kerosene from crude oil were also built; 96 dry-distillation ovens were reportedly completed in 1959.[55] Based on installation capacity, output in 1960 was approximately 300,000 tons, far below even the originally targeted 1 million tons.

The extraction rate for Maoming oil shale was 7 to 8 percent, compared with only 5.5 percent for the oil shale at Fushun. The production cost of crude oil averaged $17 to $20 (U.S.) per ton,[56] which was significantly higher than the cost of producing natural oil. For a short period after the collapse of the Great Leap Forward, construction of the Maoming shale–oil plant was suspended, but in 1965 a technically advanced oil-shale extraction plant, officially labeled the "No. 903 oil shale dry-steam furnace," was completed. This installation was followed by the completion of two more oil shale dry furnaces in 1971. As a consequence of these technical innovations, crude oil output at Maoming is reported to have advanced 31 percent in 1970,[57] 50 percent in 1971,[58] and 20 percent in 1972.[59] Maoming's crude oil output in 1974 was estimated to be 1.8 million tons. (See Table 3.9.)

Since Maoming is located very close to Chankiang, one of the major seaports in South China, excellent transportation facilities are available for exporting some of the oil to Hong Kong and Southeast Asia.

The Fushun Synthetic Oil Plants

In the 1960s China's leading shale-oil producing center was in Manchuria, centering around the city of Fushun in Liaoning province. Oil shale reserves in the Fushun area are estimated to be 5.6 billion tons. Given an oil extraction rate of 5.5 percent, the total quantity of extractable crude oil is estimated to be 300 million tons. Several shale-oil refineries and coal liquefaction plants were built by the Japanese during the 1930s and 1940s, and their capacities have been expanded since 1953. Nevertheless, after 1970 the bulk of the refining capacity was transferred to the processing of natural crude oil from Tach'ing, with a consequent sharp decline in synthetic oil production.

The Number 1 Fushun shale-oil plant was the largest shale-oil producer during the First Five-Year Plan period. Located in the western portion of Fushun, the plant was built by the Japanese in 1928.

TABLE 3.9

Estimated Crude Oil Output of the Maoming
Shale Oil Company, 1960-74
(in thousands of tons)

Year	Output
1960	300
1965	400[a]
1966	450[b]
1968	250[b]
1969	500
1970	655
1971	980
1972	1,200
1973	1,450
1974	1,800

[a]The first modern high-quality dry-steam shale-oil oven was completed in 1965. Crude output was increased by 30 percent.

[b]Output rose continuously in 1966 but suffered a severe setback during the 1967-68 disburbance caused by the Great Cultural Revolution.

Sources: (1) The figure for 1960 is from Chung-kuo Hsin-wen, March 3, 1959; (2) output for 1969 registered a 100 percent increase over the same period in 1968 and set the highest record in the history of the company, according to Nan-fang Jih-pao, February 24, 1969, indicating the deep decline of output in the 1967-68 period; (3) 1970 crude output went up 31 percent, according to Wen-hui Pao (Hong Kong), October 30, 1971, p. 3; (4) 1971 crude oil output went up 50 percent, according to New China News Agency (Peking), May 2, 1972; (5) 1972 crude oil output rose 20 percent, according to Ching-chi Tao-pao, October 1, 1972, p. 23; (6) the 1974 output was reported as 1.8 million tons, according to Fuji Jahnalu, Chugoku Keizai no Genjo to Tembo, Tokyo, 1974, p. 24.

With continuous expansion, its annual crude oil output reached 257,000 tons in 1942. The plant suffered severe damage in 1945 when the Soviet army entered Manchuria and dismantled the key equipment, and output in that year dropped to one-seventh of the original capacity. Restoration of the facilities commenced in 1948 following the Communist seizure of Manchuria. In 1949 crude output was officially given as five times that of 1946, and by 1952 it had reached the wartime peak of 257,000 tons. Several expansion projects were undertaken during the First Five-Year Plan period, and production capacity reached 675,000 tons in 1955 and 750,000 tons in 1956.[60]

In 1957, with Soviet aid, an open-cast mine was constructed at East Fushun. Installations for light oil recovery facilities were also initiated. The extraction of crude oil from shale, however, remained rather inefficient; with the antiquated round-shaped dry-distillation ovens developed by the Japanese in the 1940s, the extraction rate was low. In 1954 it took 29 tons of oil shale to extract one ton of crude. In 1959 a new type of square-shaped gas-burning oven was introduced that increased the extraction rate by more than 10 percent. The new method permits one ton of crude to be extracted from 25 tons of shale.[61] The cost of production of shale oil remained relatively high compared with that of natural oil.

The refining facilities at the Number 1 plant were obsolete, as were the production and management techniques. Its normal- and reduced-pressure distillation unit in the cracking tower, which was the only refining unit, had an annual capacity of 300,000 tons in 1960. Additional expansions increased its refining capacity to 600,000 tons by 1964. In 1970 the heating furnace and the boilers, all dating from the 1930s, also were renovated and modified. The outcome of these technical modifications increased its processing capacity to 1.8 million tons by 1973.[62]

Despite this increase in processing capacity, the production of shale oil failed to increase proportionately. The shale-oil industry has continuously encountered two problems, the shortage of shale supply and the deterioration of the quality of shale. Official reports indicate that since 1956 the shale supply has been insufficient to meet the requirements of the refineries and that the output of shale oil has declined precipitously. In 1958 the shale oil output of the Number 1 plant was reported at 430,000 tons, which amounted to only 64 percent of its capacity.[63] According to one Japanese expert who visited Fushun in 1966, the Number 1 Plant produced only 310,000 tons of crude oil in that year despite its 970,000-ton capacity.[64] This represents a 30 percent reduction in output since 1958, from which one might infer problems and/or shortages of oil shale in the 1960s. In the early 1970s most of the facilities of the Number 1 plant at Fushun were shifted to the processing of Tach'ing crude oil.

Other major oil refineries associated with shale oil or synthetic oil production are listed as follows:

The Fushun Number 2 plant, built in 1934, specialized in producing liquid fuel by the indirect liquefaction method, using bituminous coal from Fushun as raw material. In 1955 its annual crude refining capacity was 202,000 tons.[65]

The Fushun Number 3 plant, built in 1936, has an annual capacity of 250,000 tons. According to one source from Taiwan, the plant had acquired an annual processing capacity of 400,000 tons by 1973. The Number 2 plant produced high-grade gasoline and heavy and light diesel oil, and the Number 3 plant turned out jet fuel, gasoline, and ammonium sulphate.[66]

In the Kirin province of Manchuria are the Northeast Number 8 Petroleum Plant at Szeping, the Number 9 plant at Huatien, and the Number 10 plant at Mutankiang, which are also engaged in refining shale oil. Of these the Huatien plant is the largest; it is estimated to have turned out some 200,000 tons of crude oil in 1974.

Other major shale-oil refineries in the Manchuria area were transformed in order to process crude oil from Tach'ing, including the Number 4 Fushun plant built by the Japanese in 1929 (its annual processing capacity in 1974 was 2 million tons); the Number 5 plant at Chinhsi, built in 1941, with an annual processing capacity of 1 million tons; the Number 6 plant at Chinchou, built in 1939, with an annual processing capacity of half a million tons; and the Number 7 plant at Dairen, built in 1939 (its 1974 annual processing capacity was 3.5 million tons).

Estimates of the synthetic oil output of Manchuria vary widely. One Japanese estimate, apparently based on the capacity of the refineries in the ten shale-oil plants, placed output as high as 11.55 million tons in 1972,[67] ignoring the fact that 80 percent of this refining capacity is now processing natural oil. According to one source from Taiwan, output of synthetic oil in Manchuria during 1974 was approximately 1.7 million tons. The distribution is shown in Table 3.10. In view of the continuous shift of refining capacity from shale oil to natural oil and the suspected shortage in the oil shale supply, the estimate made by the Taiwan specialist appears to be closer to the mark.

INTERIOR VERSUS COASTAL OIL FIELDS

The estimates in the preceding sections reflect two fundamental changes in the Chinese petroleum industry, the changing position of interior oil fields with regard to coastal oil fields and the increased importance of natural over shale oil.

TABLE 3.10

Distribution of Synthetic Oil Output in Manchuria Area, 1974
(in thousands of tons)

Plant	Location	Production
Northeast Number 1	Fushun, Liaoning	500,000
Northeast Number 2	Fushun, Liaoning	400,000
Northeast Number 3	Fushun, Liaoning	400,000
Northeast Number 8	Szeping, Kirin	20,000
Northeast Number 9	Huatien, Kirin	200,000
Northeast Number 10	Chiao-ho, Kirin	150,000
Mutankiang Refinery	Mutan, Heilungkiang	20,000
Ch'engte Refinery	Ch'engte, Hopei*	30,000
Total		1,720,000

*Ch'engte was a part of Manchuria but has since come under the jurisdiction of Hopei province.

Source: Chuan Wei, "A Critical Review on Recent Development of Petroleum Industry on Chinese Mainland," Ta-lu Ching-chi yen-chiu 7, no. 1 (1975): 50-51.

The changes in the regional distribution of crude oil production are clearly illustrated in Table 3.11 and Map 3.3. In 1960, two-thirds of China's crude oil was produced by oil fields in the north-west and southwest regions; indeed, two major oil fields in the west, Yumen and Karamai, turned out one-half of China's total crude oil. The situation underwent a drastic transformation following the opening of Tach'ing in Manchuria, and in 1965 the crude oil output of the coastal oil fields accounted for 57 percent of the total. By 1970, two-thirds of China's crude oil came from these coastal oil fields. Within 14 years the relative production importance of the two areas had been reversed.

The relative decline of the contribution of shale oil to the total crude supply was equally conspicuous. In 1952 shale oil accounted for 55 percent of China's crude output; the proportion had shrunk to 25 percent by 1960; 13 percent by 1965; 9 percent by 1970; and 7 percent by 1974.

These changes stem mainly from cost considerations. The cost of extracting crude from oil shale has always been relatively high in comparison to natural oil production; for years the cost of the shale oil produced in Fushun exceeded its output value.[68] The

TABLE 3.11

Regional Distribution of Crude Oil Output, 1960–74
(in percent)

Oil Field	1960	1965	1970	1974
Northeast				
Tach'ing	9.0	38.0	40.0	37.0
Fushun and other				
shale refineries	19.0	9.0	7.0	4.0
North China				
Takang	--	--	4.0	7.0
East China				
Shengli	--	6.0	13.0	19.0
Others	--	--	--	2.0
South China				
Number 70 oil field	--	--	1.0	2.0
Maoming	5.0	4.0	2.0	3.0
Total for coastal areas	33.0	57.0	67.0	74.0
Northwest				
Yumen	26.0	11.0	8.0	4.0
Karamai	23.0	24.0	11.0	11.0
Tushantzu	1.0	1.0	1.0	1.0
Tsaidam	10.0	2.0	7.0	6.0
Yench'ang	1.0	1.0	1.0	1.0
Southwest				
Szechwan	6.0	4.0	5.0	3.0
Total for interior areas	67.0	43.0	33.0	26.0
Total for all identified				
oil fields	100.0	100.0	100.0	100.0

MAP 3.3

Regional Distribution of Crude Oil Production, 1960–74

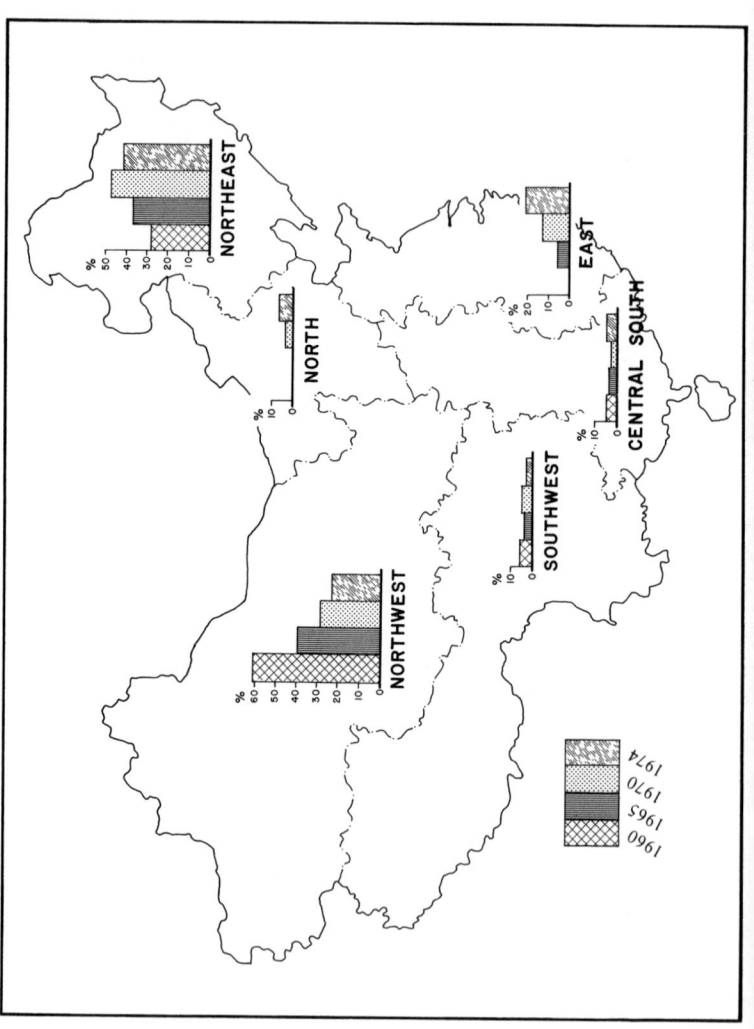

rationale for the continuation of shale-oil production in the 1960s was
the low cost of transportation. Since low-cost natural oil is now
available in large quantities in the coastal areas, the substitution of
natural for synthetic oil production is based on cogent decisions with
respect to production and transportation cost data. The extremely
high overhead costs involved in the exploration of the interior oil
fields has discouraged their further development. On the other hand,
as mentioned above, the total profit of the coastal Tach'ing field over
the past 14 years was reportedly equivalent to 11 times its capital
investment. Unless China is willing to allocate the capital-intensive
investment necessary to improve the transportation network into the
interior areas, the northwest oil fields will probably not witness substantial new development during the late 1970s.

EVALUATION OF THE ESTIMATES

As noted at the outset of this chapter, due to the paucity of official statistics, the derivation of time series of crude output for individual oil fields has to depend on a wide variety of sources. The
result of the derivation must be cross-examined for internal consistency. First, the sum total of the individual oil field output must be
compatible with the estimated national total. Second, the regional
distribution of crude output must be congruous to conditions such as
refining capacity and transportation facilities.

Of the existing studies on the Chinese petroleum industry, very
few offered estimates on crude output for individual oil fields. Table
3.12 summarizes three estimates made by experts in Taiwan (Chuan
Wei), Japan (Ohno Hideo), and the United States (Bobby A. Williams).
Comparing these estimates for different years with my estimates for
the corresponding years, some evaluation of the internal consistency
of different estimates can be made.

Ohno's estimate of crude output for 1972 was approximately 20
percent less than my estimate, which was based on the secret document of the Chinese government. The main reason for Ohno's underestimation stems from two factors: first, he did not include crude
output from many new oil fields in the east coast such as Takang,
Itu, "913," "70," and "57"; second, he did not leave room for some
oil fields which have been developed but still not announced. However,
in terms of regional distribution Ohno's estimate coincided with mine,
being 27 percent for the interior oil fields and 73 percent for the
coastal oil fields.

My estimated total for 1973 is 50 percent higher than Chuan
Wei's. For crude oil output of the interior oil fields Chuan Wei's
figure, 12.1 million tons, comes very close to my own figure, 12.9

TABLE 3.12

Comparisons of Four Estimates of Crude Oil Output of Major Oil Fields, 1972-74
(in millions of tons)

Oil Field	1972		1973		1974	
	Author	Ohno (Japan)	Author	Chuan-Wei (Taiwan)	Author	Williams (U.S.)
Yumen	2.30	2.20	2.50	3.50	2.65	0.71
Kuramai and Tushantzu	4.50	4.50	5.00	5.10	6.00	1.04
Yen-ch'ang	0.20	1.50	0.40	0.20	0.40	0.12
Tsaidam	2.80	0.90	3.00	1.10	3.50	0.53
Szechwan	1.80	0.60	2.00	2.20	2.20	2.00
Total, interior	11.60 (27%)	9.70 (27%)	12.90 (26%)	12.10 (35%)	14.75 (26%)	4.40 (10%)
Tach'ing	15.30	15.10	17.00	14.30	20.74	19.40
Shengli	7.20	8.00	8.80	4.00	10.21	11.00
Takang	1.90	--	3.30	2.00	4.12	3.74
Shale oil	3.50	4.10	4.00	2.60	4.00	3.00
Others	2.70	--	4.00	--	4.80	6.00
Total, coastal	30.60 (73%)	27.10 (73%)	37.10 (74%)	22.90 (65%)	43.87 (74%)	43.14 (90%)
Unknown	2.80	--	3.00	--	4.38	19.56
National total	45.00	36.90	53.00	35.00	63.00	65.30

Note: Figures in percentage show the distribution between interior and coastal oil fields. Estimates from different authors are the latest estimates each author provided. Ohno's latest year is 1972, Chuan-Wei's latest year is 1973, and Williams' latest year is 1974.

Sources: Author's estimates from Table 2.8. Chuan-Wei's estimates from "A Critical Review on Recent Development of Petroleum Industry on Chinese Mainland," Ta-lu Ching-chi Yen-chiu 7, no. 1 (January 1975): 41–60. Ohno's estimate from Ohno Hideo, "Transition Period of Chinese Petroleum Industry," Chugoku Kogyo Tsushin (Tokyo), February 1973, Table 2, p. 9. Williams' estimate from Bobby Williams, "The Chinese Petroleum Industry: Growth and Prospects," in U.S. Congress, Joint Economic Committee, China: A Reassessment of the Economy (July 1975), p. 233.

million tons. The discrepancy of 18 million tons between our total estimates arises from the fact that Chuan Wei's estimate for crude output in the new oil fields on the east coast is considerably below the figures available in Japan and the West. Chuan Wei's estimate for crude output in Tach'ing, Shengli, and Takang amounted to only 20 million tons, 9 million tons less than my estimate. Moreover, he did not take into account crude output from those newly developed oil fields. Consequently, the national total crude output for 1973 estimated by Chuan Wei added up to only 35 million tons, a figure far below the 50 million tons plus disclosed by the Chinese government.

My estimate for 1974 was in agreement with Bobby Williams' for the new oil fields in coastal areas, the estimates being 43.87 million tons and 43.14 million tons respectively, and for the national total, 63 million and 65 million tons respectively. The most conspicuous difference between my estimate and Williams' lies in our figures for crude output of interior oil fields. Williams' estimate for the five oil fields in the interior area--Yumen, Karamai, Yench'ang, Tsaidam, and Szechwan--amounted to only 4.4 million tons, a figure even less than the annual output for Karamai alone as estimated by most experts in Taiwan, Hong Kong, and Japan. As a consequence, Williams' estimate shows 19.56 million tons of crude oil, or 30 percent of his estimated national total, whose sources are classified as "unknown." Since Williams already has fully taken into account crude oil output in all known new oil fields including Hupch, Fu-yu, P'an Shan, and I-tu, it is difficult to believe that there is some "unknown" oil field which produces an annual output of 19.56 million tons, an output the size of the much publicized Tach'ing.

Moreover, following Williams' estimate, the regional distribution of known oil field output becomes only 10 percent for the interior but 90 percent for the coastal, a proportion which is not in line with the regional distribution of the refining capacity. As shown in Table 4.4, the regional distribution of the refining capacity in 1973 was 78 percent for the coastal areas and 22 percent for the interior. This figure basically dovetails with my estimate for crude oil in that year, which is 74 percent for the coastal and 26 percent for the interior. If Williams' estimate for interior crude oil output is accepted, it follows that a part of the coastal oil, amounting to some 10 million tons, had been shipped to the interior for refining. The transportation facilities in China simply exclude this possibility.

NOTES

1. "Karamai, Newest and Biggest Oilfield," China Reconstructs, November 1956, p. 8.

2. Ku Lei, "Tsaidam," China Reconstructs, April 1957, p. 2.

3. The former figures are estimated by Ch'en Cheng-siang, "The Petroleum Resources of China and Their Development," in Research Report (Hong Kong: The Chinese University of Hong Kong, 1966), p. 13, Table 2.

4. Ibid., p. 10.

5. Far Eastern Economic Review, April 2, 1959, pp. 470-71.

6. China Reconstructs, December 1973, p. 32.

7. China Pictorial, no. 2 (1974), pp. 5-6.

8. Jen-min Jih-pao, August 30, 1974.

9. New China News Agency (Lanchow), September 10, 1974.

10. Y. I. Berizina, Fuel and Power Base of the Chinese People's Republic (Moscow, 1959); Joint Publications Research Service, no. 3789, August 31, 1960.

11. Ch'en Cheng-siang, op. cit., p. 10.

12. "Karamai, Newest and Biggest Oilfield," op. cit., p. 8.

13. Berizina, op. cit., p. 49.

14. Urumchi-Sinkiang Radio, December 12, 1973.

15. Joint Publications Research Service, no. 61,534, p. 15.

16. Ku Lei, op. cit., p. 2.

17. China News Analysis, no. 406 (February 2, 1962), p. 6.

18. Ibid.

19. Shih-yu K'an-t'an, no. 4 (1960), pp. 12-13.

20. China Reconstructs, January 1973, p. 27; Ta-kung Pao, Hong Kong, July 7, 1974.

21. Joint Publications Research Service, no. 55,638, p. 13.

22. China Reconstructs, January 1973, p. 23.

23. "A Visit to the Tsaidam," China Pictorial, no. 6 (1974), pp. 32-34.

24. China Quarterly, December 1974, pp. 834-35.

25. Chao Yu-sheng, "The Tach'ing Oilfield," in Collected Documents of the First Sino-American Conference on Mainland China (Taipei: Institute of International Relations, 1971), p. 797.

26. Chung-kuo Hsin-wen, December 30, 1972, p. 12.

27. Peking Review, June 7, 1974, p. 16.

28. Ibid., p. 17.

29. Wilfred Burchett, "China Taps Tach'ing Potential," Far Eastern Economic Review, January 14, 1974, p. 47.

30. Peking Review, June 7, 1974, p. 17.

31. Peking Review, October 11, 1974, pp. 16-17.

32. Jen-min Jih-pao, October 24, 1974.

33. Peking Review, January 4, 1974, p. 5.

34. Chang Chun, "How China Developed Her Oil Industry," China Reconstructs, October 1974, p. 7.

35. Peking Review, October 11, 1974, p. 16.

36. Chuan Wei, "A Critical Review on Recent Developments of Petroleum Industry on Chinese Mainland," in Ta-lu Ching-chi Yen-chiu, January 1957, p. 44; Foreign Broadcast Information Service, Daily Report (December 10, 1974), p. G-7.

37. The Oil and Gas Journal, August 25, 1975, p. 23.

38. The China Quarterly, December 1974, p. 835.

39. Hung-ch'i, August 1974.

40. China Reconstructs, October 1974, pp. 8-14.

41. Ibid.

42. New China News Agency, October 4, 1974.

43. Yoshio Koide, "China's Crude Oil Production," Pacific Community 5, no. 3 (April 1974): 462-70.

44. John Cranfield, "Mainland China Gearing Up to Boost Oil Exports," The Oil and Gas Journal, August 11, 1975, p. 22; Paul H. Fan, "Chinese Oil-Industry Image Changing," op. cit., pp. 110-12.

45. Wall Street Journal, August 10, 1974, p. 1.

46. Shih-yu K'an-t'an, no. 8 (April 14, 1958), p. 4.

47. A. A. Meyerhoff, "Development in Mainland China, 1949-1968," The American Association of Petroleum Geologists Bulletin 54, no. 8 (August 1970), p. 1579.

48. Ibid.

49. Ho Ko-jen, "The Development in Red China's Petroleum Industry," Fei-ch'ing Yueh-pao (Taipei), March 1, 1968, pp. 52-65.

50. New China News Agency (Ch'engtu), January 13, 1972.

51. United Nations Economic Commission for Asia and the Far East, Economic Survey of Asia and the Far East, 1949 (Lake Success, N.Y.: the UN, 1950), p. 341.

52. New China News Agency (Peking), September 27, 1958.

53. China Petroleum Corporation, Ta-lu shih-yu kung-yeh hsian-shih [The Present Status of the Petroleum Industry in Mainland China] (Taipei: the Corporation, 1968), pp. 61-63.

54. Chung-kuo Hsin-wen, March 3, 1959.

55. Ibid.

56. Togawa Toru, "China's Petroleum Industry," Chugoku-Keizai-Kenkyu-Geppo, July 1973, pp. 35-36.

57. Wen-hui Pao (Hong Kong), October 30, 1971, p. 3.

58. New China News Agency (Peking), May 2, 1972.

59. Ching-chi Tao-pao, October 1, 1972, p. 23.

60. Shih-yu Kung-yeh T'ung-hsun, no. 18 (1957), pp. 38-41; Kwang-ming Jih-pao, July 6, 1957, and September 6, 1957.

61. Ku Ching-hsin, "Shale Oil Industry of Our Country Achieved Rapid Progress," Jen-min Jih-pao, May 1, 1959.

62. New China News Agency (Shenyang), March 3, 1974.

63. K. C. Yeh, Communist China's Petroleum Situation (Santa Monica, Calif.: Rand Corporation, 1962), p. 35.

64. Togowa Toru, op. cit., pp. 33-34.

65. Kwang-ming Jih-pao, July 6, 1957.

66. Chuan Wei, op. cit., pp. 50-51.

67. Fuji Jahnalu, Chugoku Keizai no Genjo to Tembo (Tokyo, 1974), p. 21.

68. New China News Agency, March 3, 1974.

4

RESERVES, REFINERIES,
AND TRANSPORTATION

The development of a given petroleum industry is conditioned by four major factors: (1) the existence of sufficient reserves; (2) the capacity for drilling and extracting; (3) the capacity for refining the crude oil into marketable products; and (4) the existence of adequate facilities to transport the petroleum from oil field to refinery and from refinery to market. In the preceding Chapters 2 and 3 the focus was on China's capacity for drilling and producing crude oil; in this chapter, attention will be centered on the other three factors.

PETROLEUM RESERVES

Reliable data on petroleum reserves are relatively scanty for most countries and even scarcer for China. Estimating proven reserves involves the assessment of vast areas of sedimentary basins of little-known geology that have never been drilled. Even in the United States, where petroleum geologists have done more extensive and intensive studies than elsewhere in the world, estimates of the country's oil-producing potential differ.[1]

The concept of oil reserves is understood in various ways. In the West, oil reserves are generally divided into three categories: proven reserves, probable reserves, and possible reserves. Proven reserves refers to recoverable oil proven by drilling in an oil field, which may be estimated with a reasonable accuracy. Probable reserves refers to the reserves judged to be contained in extensions of the proven oil-bearing area in an oil field under development. Possible reserves refers to the oil that is expected to be found in addition to the proven and probable reserves in areas not yet under development.[2] Development drilling of an oil field must be completed before one can speak of its proven reserves. In oil fields in an advanced stage of development there are mostly proven and probable reserves. Possible reserve estimates are made on the basis of the geological aspects of the areas that have not yet been drilled; needless to say, these estimates are highly speculative. In many countries the proven reserves are still only a small proportion of the probable reserves. In 1945, in the case of the United States, only 8 percent of the total probable reserves were counted as proven reserves.[3]

Reserves are measured far less conservatively in countries outside the United States, and in the case of China, the very limited data on oil reserves are even more difficult to assess. In the first place, there was no extensive geological survey before 1949, and in the subsequent 11 years, only a small portion of the Chinese mainland was explored carefully.[4] Although continuous prospecting since the early 1960s has led to the discovery of several promising oil fields, no authoritative estimate on China's oil reserves is available.

Data interpretation is made further difficult by the fact that Chinese official reports have followed the Soviet practice of using the term "A + B class" for proved and probable reserves and "C class" for possible reserves. *

In 1949 the probable oil reserves in China were officially estimated at 200 million tons.[5] The special goal of proven reserves containing 55 million tons of crude oil was set for the First Five-Year Plan (1953-57); as a result of the extensive prospecting in 1956,

---

*The Chinese A class of reserves is subdivided into two categories: (1) "A-1 class," which corresponds to proved reserves in the West, refers to oil deposits that have been verified and that are extractable with a producing well; (2) "A-2 class" is similar to the "known reserve" in the West and refers to an area that is bounded by two or more wells with an industrial oil flow. The Chinese "B class" refers to any verified industrial oil field within a known oil field of uncertain boundary. The "C-1 class" is estimated on the basis of the lowest physical properties, stability, and other information about an oil formation that is either newly discovered or discovered by a newly drilled well with an industrial oilflow in a new prospecting region, and the "C-2 class" refers to the estimated oil reserve in a possible oil field by geological survey. (K. C. Yeh, Communist China's Petroleum Situation [Santa Monica, Calif.: Rand Corporation, 1962], p. 28 Note; Shih-yu K'an-t'an, no. 7 [1958], p. 17.)

this target is believed to have been overfulfilled. In 1957, the last
year of the first plan, an estimate of possible reserves (or potential
reserves) of 1.7 billion tons was also given by official sources. [6]
The quantity of proven reserves was approximately 100 million tons.
When the Second Five-Year Plan started in 1958, proven reserves
were planned to reach 150 to 200 million tons by 1962, [7] a figure that
was less than one-eighth of the estimated possible reserves in 1957.

The Great Leap Forward geological prospecting and exploration
program in 1958-59 significantly enlarged the proven reserves. In
early 1960 an official report stated that the proven reserves drilled
during the 1958-59 period surpassed the 200-million-ton target set
for 1962. It was also stated that the area of sedimentary strata containing crude oil and natural gas had proved to be 70 percent larger
than the area known in 1957. [8] Total potential reserves in 1960 may
amount to 2.9 billion tons, in contrast to the 1.7 billion tons estimated for 1957.

Since 1960 there has been a total blackout of official petroleum
reserve data. In 1964 a Chinese technician, previously graduated
from the Peking Petroleum Institute and associated with the Ministry
of Petroleum Industry of the People's Republic of China, fled from
China to Taiwan, bringing with him a substantial amount of data on
the Chinese petroleum industry. Armed with this information, he
compiled a table showing the known reserves for the major oil fields
in China. [9] His estimate has been widely accepted by experts in
Taiwan, Japan, and the United States. A comparison of his estimates of the oil reserves in 1966 and the estimates made by other
authors for the earlier years is presented in Table 4.1.

In the West, the most authoritative estimates of China's oil
reserves are made by A. A. Meyerhoff. According to Meyerhoff,
at the beginning of 1969, China's proved-plus-probable-plus-potential-
and-possible reserves of natural crude oil totaled 2.68 billion metric
tons. Of this total, he considered proved reserves to be 182 million
tons and probable reserves to be 777 million tons, for a total of 959
million tons in the A and B classes, [10] an estimate very close to the
1 billion tons estimated in Table 4.1.

Official Chinese reports indicate that since 1966 large-scale
oil prospecting has been going on in many parts of China and that
many new oil and gas fields have been discovered. Reference is also
made to successful experimental prospecting for oil hidden in the
continental shelf. The most promising discovery, which has drawn
worldwide attention, is the proclaimed offshore reserves. These
deposits in the continental shelf stretch from the Yellow Sea between
Korea and the Shantung Peninsula to as far as Hsi-sha (Paracel) and
Nan-sha (Spratley), islands in the South China Sea. If initial prospects are proved, offshore oil could represent a prodigious addition
to Chinese petroleum capacity.

TABLE 4.1

Estimated Natural Petroleum Reserves, 1966

| Oil Field | Various Years[a] | | | | By the End of 1966[c] | | |
	Year	Grade	Area[b] (in square kilometers)	Reserves (in millions of tons)	Grade	Area[b] (in square kilometers)	Reserves (in millions of tons)
Sinkiang							
Tushantzu	1950	C	3,900	400	B	130	20
Karamai	--	--	--	--	A, B	169	100
	--	--	--	--	C	4,500	2,100
Kansu							
Yumen	1956	A, B, C	7,000	300	A, B	90	280
					C	7,000	NA
Ya-erh-hsia	--	--	--	--	A	20	70
	--	--	--	--	C	500	600
Tsaidam							
Lenghu	1953	C	10,700	1,400	A, B	100	130
	--	--	--	--	C	380	NA
Others	--	--	--	--	A, B	962	NA
	--	--	--	--	B, C	2,000	1,000
Szechwan							
Central	1958	B, C	1,000	--	A, B	1,200	80
	--	--	--	--	C	10,000	NA
Southern							
(natural gas)	1957	B, C	2,300	100 billion cubic meters	B	700	500 billion M^a
					C	15,000	
Sungliao Plain							
Tach'ing	--	--	--	--	A, B	1,000	120
	--	--	--	--	C	24,000	900[d]
East China							
Shenghi	--	--	--	--	B	30	40
	--	--	--	--	C	2,500	NA
Shensi							
Yench'ang	1951	B	6	20	B	12	50
Southwest							
Tibet Kueichow and Yunan	--	--	--	--	C	9,500	2,000
Other (shallow oil)	1958	B	1,020	--	B	209	34
Total natural oil reserves					A, B	4,622	1,000
	1957	C	23,900	1,700[b]	C	75,380	6,000

NA = Not available.

[a]See Source 1. [c]See Source 3.
[b]See Source 2. [d]See Source 4.

Sources: (1) Report on "China's Special Industries," Tekko-Kaigai-Shijyo-Chosa-Kynkai-Kan [Journal Investigation Committee for Overseas Market of Iron and Steel], Tokyo, March 1972, p. 17; (2) Ho K'o-jen, "Peiping's Petroleum Industry," Issues and Studies 4, no. 11 (August 1968): 28-29; (3) Brian Heenan, "China Petroleum Industry," Far Eastern Economic Review 49, no. 13 (September 23, 1965): 567. (4) The figures for Sungliao Plain were revised upward in Li Tao-kwei, "Energy Crisis in the World and the Role May Be Played by the Chinese Communists," Ta-lu Ching-chi Yen-chiu [Studies on the Economy of Mainland China] (Taipei: 1973), 5, no. 4: 7-8.

Although little information is available on China's offshore reserves of natural crude oil, preliminary geophysical data gathered by Western scientists in the late 1960s show great potential in the Chinese continental shelf. A study by Wageman, Hilde, and Emery concluded that "the continental shelf between Taiwan and Japan may be one of the most prolific oil reservoirs in the world."[11] A Chinese journal, K'o-hsueh Shih-yen (Scientific Experiment), echoed this optimistic view in 1973, stating:

> The offshore sedimentary basins lying on the continental shelves between the T-iao-yu (Shenkaku) Islands at the farther end of the East China Sea and the coast of the China Mainland are considered to be most favorable to oil prospecting and exploitation. Without any doubt, this could lead to the discovery of a significant number of oil and gas fields in the future.[12]

With this in mind, recent estimates of Chinese possible reserves have been revised substantially upward, generally ranging from 10 to 50 billion tons. One estimate made in West Germany put Chinese oil reserves at 8 to 10 billion tons;[13] another, by an American executive who visited China in the spring of 1974, placed the total reserves at between 40 and 45 billion tons. This wide discrepancy can probably be attributed to differing frames of reference: while the figure of 8 to 10 billion tons may refer to "probable reserves," the figure of 40 to 45 billion tons more probably denotes "possible reserves."

More recently, Chinese official reports claimed that "many high-yield wells were drilled in China during 1974. Test drilling was very successful in some key areas under exploration. Oil-bearing structures in some other areas were confirmed to be larger than heretofore known, and promising oil and gas reserves were found in other areas."[14] This statement tends to reinforce speculation that China may soon have the third-largest oil reserves in the world.

In recent years sinologists have estimated that China's possible reserves, added to the recently discovered offshore deposits, will aggregate 30 billion tons, or 146 billion barrels. Of this total, only 3 billion tons, or 22 billion barrels, can be counted as A + B class, which is to say proven and probable, reserves.[15]

In the United States, as crude output has increased, the ratio of proven reserves to annual production has gradually declined, indicating that the oil reserve depletion has exceeded discovery of new reserves. The ratio fell from 16.3 years in 1920 to 15.1 years in

1930, 14.1 years in 1940, 13.5 years in 1950, 12.8 years in 1960, and 8.9 years in 1970.[16] Will China be confronted with a similar pattern? How long will the Chinese oil reserves last?

If we assume that in 1971 China's A + B class of reserves was 3 billion tons, approximately equal to the proven reserves of the United States in 1945, the reserve-output ratio prior to 1975 will be more than 30 years. (See Table 4.2.) However, as China's output of crude oil continues to increase, the probable reserve may be rapidly depleted. If one assumes no concurrent increase in probable reserves and a 20 percent annual growth rate of crude oil production in 1975-77; a 17 percent growth rate in 1978-80; 15 percent in 1981-83; and 12 percent in 1984-85, the reserves-production ratio will drop to only two years by 1985. Maintaining a reserves-production ratio of 30, given these assumed production increases, would require a total of 8 billion tons of new probable reserves. In short, by 1985 China must increase its probable reserves by 160 percent more than has been discovered since the early 1950s (3 billion tons) if the reserves-production ratio of 30 is to be maintained. To accomplish this, China could drill more wells into and around known reservoirs or search for a new oil pool in other areas. All such efforts would require substantial amounts of exploration investment. Whether the Chinese petroleum industry can maintain such a high reserves-production ratio is contingent on the amount of exploration investment as well as on technological improvements.

In addition to her reserves of liquid oil, China possesses enormous deposits of shale and coal from which synthetic oil may be processed. Oil-bearing shale deposits were officially estimated at 50 billion tons and coal reserves at 1,500 billion tons in 1958.[17] At the present extraction rate of 5 percent (1 ton of petroleum from 20 tons of shale), the 50 billion tons of oil shale might add an additional 2.5 billion tons of oil to the existing reserves of liquid oil. Of the huge coal reserves, 10 billion tons have been verified as recoverable. Much of this is bituminous coal, a large proportion of which is thought to have an oil content as high as 46 percent.[18] Given these resources, one might surmise that in the foreseeable future China may have adequate oil reserves for her petroleum industry.

REFINERY CAPACITY

As in the case of crude production, refinery capacity in China was concentrated in Manchuria during the initial period of development. The facilities, built by the Japanese in the 1930s, were primarily intended for shale-oil refining, and very few were designed for processing natural crude oil.

TABLE 4.2

Projected Crude Oil Reserves Requirements for 1971-85
(in millions of tons)

Year	Output[a]	Cumulative Crude Output	Remaining Reserves[b]	Reserves-Production Ratio[c] (in years)	Reserves Required[d]	New Reserves Needed[e]
1971	38	38	2,962	78		
1972	45	83	2,917	65		
1973	53	136	2,864	54		
1974	63	199	2,801	44		
1975	76	275	2,725	36		
1976	91	366	2,634	29	2,730	
1977	109	475	2,525	23	3,270	540
1978	128	603	2,397	19	3,840	570
1979	150	753	2,247	15	4,500	660
1980	176	929	2,071	12	5,280	780
1981	202	1,131	1,859	9	6,060	780
1982	232	1,363	1,637	6	6,960	900
1983	267	1,630	1,370	5	8,010	1,050
1984	299	1,929	1,071	4	8,970	960
1985	335	2,264	736	2	10,650	1,680

[a]See Source 1.

[b]Three billion tons (assumed as probable reserves in 1971), less output. See Source 2.

[c]Remaining reserves divided by output.

[d]Output times 30, assuming the necessity of a reserves-production ratio of 30.

[e]The difference in reserves required between a given year and the year preceding.

Sources: (1) Output figures for 1971-74 are compiled from the various estimates listed in Tables 2.4 and 2.5; output figures for 1975-77 are derived from these, assuming a growth rate of 20 percent annually; for 1977-80, assuming 17 percent; for 1981-83, assuming 15 percent; for 1984-85, assuming 12 percent. Sources are found in Table 2.13.
    (2) The figure of 3 billion tons (A + B reserves) is from Chien Yuan-heng, "Chinese Communist Petroleum Industry as Seen from the World Wide Energy Crisis," Studies on Chinese Communism Monthly (Taipei, May 10, 1974), p. 61.

Most of the shale-oil refineries built by the Japanese suffered devastating damage following the termination of the Sino-Japanese War in 1945, when the Russians removed most of the machinery and equipment. In 1949 the total oil refinery capacity in China dropped to only 160,000 tons, less than one-third of the 500,000 tons of the prewar peak.

Intensive efforts were made in the 1950-52 period to restore the refinery capacity in Manchuria. The Number 1 Fushun Refinery, the largest synthetic plant in China, was producing at only 30 percent of its original capacity of 300,000 tons by 1949, while still undergoing reconstruction. Efforts were also made to rehabilitate the other six refineries in the Manchuria area, as well as the small old refinery at the Yumen oil field in Kansu. By 1952 China's refining capacity approached 1 million tons.

The First Five-Year Plan goal was aimed at renovating several of the old refineries and constructing one large modern refinery. The Number 1 Fushun Refinery regained its prewar output level in 1952. Through technical innovations, its capacity in 1956 substantially surpassed the 1943 level. The Yumen refinery also underwent expansion and renovation. The semicracking equipment that had previously been installed was replaced by Soviet-made cracking equipment that greatly increased its processing capacity.

The most significant development in this period, however, was the construction of the modern Lanchow Refinery at Kansu, with complete Soviet equipment and technology. The plant consists of 16 main production units, including a catalytic cracking unit for making aviation gasoline, a propane deasphalting unit for producing lubricants, and a unit for processing natural gas. The new enterprise, which commenced construction in 1956 and began partial production in 1959, has significantly augmented Chinese refining capacity. Chen Chengsiang estimated that by the end of 1957 China's total refining capacity had reached 2.2 million tons and exceeded her crude oil output by .7 million tons.[19]

In 1958 the government announced a target goal of 5.5 million tons of refining capacity by the end of 1961.[20] During this period two small refineries, in Shanghai and Nanking on the eastern coast, were vigorously expanded and a dozen new small- and medium-sized refineries were erected at the newly discovered oil fields in Tsaidam and Szechwan. A modern large synthetic oil refinery was also constructed at Maoming, in Kwangtung province in Southern China.

In Shanghai a small refinery left intact by the former government underwent large-scale expansion. Its crude-oil processing capacity in 1958 was reported to have increased ten times over what it had been in 1954.[21]

In Nanking the refinery was also enlarged, raising its capacity from less than 100,000 tons to 1 million tons in 1960.

In the Tsaidam Basin five small-scale refineries were constructed. The principal one was located at Lenghu, with a capacity of 100,000 tons, and subsequently was expanded to 300,000 tons. Another was built at Yuchuntzu with an annual capacity of 350,000 tons.

In the Nanch'ung area of Central Szechwan, 11 small-sized refineries were built, with a total capacity of half a million tons a year.[22] Thus, by expansion of existing facilities and new construction, the total capacity of China's refineries reached 7 million tons by the end of 1962. This refining capacity slightly exceeded the crude oil output for that year and greatly surpassed the targeted goal of 5.5 million tons that had been set for the end of 1961.

However, the great expansion of refining capacity came after 1962, paralleling the surging crude oil production. According to official reports, the refining capacity increase during 1960-69 was eight times the increase of 1950-59.[23] During 1950-59 China's refining capacity increased from .5 million tons to 4 million tons, a net gain of 3.5 million tons. The net addition to refining capacity during the 1960-69 period thus amounted to 28 million tons (8 x 3.5 = 28). By 1970 China's refinery capacity reached 28.5 million tons, thereby keeping pace with the growth of her crude oil output. Subsequently the gap between refining capacity and crude oil output has widened. In 1971, when crude output expanded 28 percent, refining capacity increased by only 16 percent; in the following year, while crude output rose 16 percent, refining capacity advanced only 5 percent. Thus Chinese refineries could only process 80 percent of the crude oil produced in 1973 and about 75 percent in 1974. (See Table 4.3.)

Nevertheless, the eightfold increase in refining capacity during the 1960-69 period significantly improved the structure and geographic distribution of China's refining industry. (See Table 4.4.)

In the first place, the average size of refineries was notably enlarged. In the early stage of development, 1950-57, no refinery in China had an annual capacity exceeding 500,000 tons, but since the early 1960s China has built and expanded several refineries having a capacity of 4 million tons a year, which produced considerable economies of scale. Experiences in Western Europe, the Caribbean, and Japan during the 1965-66 period evidence that as the operations increased in scale from 1 million tons of crude oil a year to 4 million, the capital investment required could be reduced up to 40 percent.[24] By 1974, six known refineries in China possessed a processing capacity of 3.5 million tons of crude oil per year. The major refineries are listed as follows:

TABLE 4.3

Estimated Oil Refining Capacity, Various Years

Year	Capacity (in thousands of tons)	Index 1952 = 100	Index 1957 = 100
1949	160	16	7
1952	1,000[a]	100	45
1957	2,200	220	100
1959	4,000	400	182
1960	5,500	550	250
1964	16,000[b]	1,600	727
1970	28,500	2,850	1,295
1971	33,000[c]	3,300	1,500
1972	38,000[d]	3,800	1,727
1973	42,000[e]	4,200	1,909
1974	47,500[f]	4,750	2,159

[a] According to the First Five-Year Plan, capacity in 1957 was to be increased by 1.5 times over that of 1952. Since 1957 is estimated at 2.6 million tons, capacity should have been about 1 million tons in 1952.

[b] 1974 capacity was reportedly three times that of 1964. See Source 3.

[c] 1971 capacity went up 16 percent. See Source 5.

[d] 1972 capacity went up 5 percent. See Source 6.

[e] A 10 percent increase is assumed for 1973. See Source 3.

[f] 1974 capacity went up 13 percent. See Source 3.

Sources: (1) Figures for 1949 and 1957 are from Ch'en Cheng-siang, "The Petroleum Resources of China and Their Development," in Research Report (Hong Kong: The Chinese University of Hong Kong, 1966), pp. 15-16; (2) figures for 1959 and 1960 are from Shih-yu Lien-chih, no. 4 (1958), pp. 5-9 (these are planned figures); (3) figures for 1969 and 1974 are based on Hsueh Shih-shih, no. 1 (1975), p. 6; (4) figures for 1970 are derived from the information in Chung-kuo Hsin-wen, September 2, 1973, p. 4; (5) figures for 1971 are from New China News Agency (Peking), January 9, 1972; (6) figures for 1972 are based on Ta-Kung-Pao (Hong Kong), December 29, 1972.

TABLE 4.4

Estimated Capacity of Major Refineries, 1960–73

(in millions of tons)

Refinery	Location	Capacity			
		1960	1964	1970	1973
Lanchow	Lanchow, Kansu	1.00	2.00	4.00	4.50
Tach'ing	Tach'ing, Heilungkiang	--	1.00	4.10	5.00
Shanghai	Shanghai	0.20	2.50	3.50	4.00
Yumen	Yumen, Kansu	0.35	1.00	1.20	1.50
Nanking	Nanking, Kiangsu	1.00	2.00	2.50	3.50
Dairen Number 7	Dairen, Liaoning	0.50	1.00	1.20	3.50
Fushun Number 1	Fushun, Liaoning	0.50	1.20	1.20	1.50
Fushun Number 4	Fushun, Liaoning	0.20	0.30	1.00	2.00
Tushantzu	Tushantzu, Sinkiang	0.40	1.00	2.00	2.00
Anshan	Anshan, Liaoning	--	--	--	2.50
Hangchow	Hangchow, Chekiang	--	--	--	1.00
Nanch'ung	Nanch'ung, Szechwan	0.30	0.50	0.50	0.60
Lenghu	Tsaidam, Tsinghai	0.30	0.30	0.70	0.90
Peking General Petrochemical	Peking, Hopei	--	--	2.50	4.00
Maoming	Maoming, Kwangtung	--	2.00	2.00	2.50
Chin-hsi Number 5	Chinhsi, Liaoning	0.50	0.50	1.00	1.00
Tientsin	Tientsin, Hopei	--	0.20	0.50	1.00
Other		0.25	0.50	0.60	1.00
Total capacity		5.50	16.00	28.50	42.00

Sources: (1) Figures for 1960 are based on K. C. Yeh, Communist China's Petroleum Situation (Santa Monica, Calif.: Rand Corporation, 1962), Appendix; (2) figures for 1964 are based on Ch'en Cheng-siang, "The Petroleum Resources of China and Their Development," in Research Report (Hong Kong: The Chinese University of Hong Kong, 1966), p. 13, Table 2; (3) figures for 1970 are based on Kambara Tatsu, "Petroleum Industry in China," Sekiyu-no-Kaihatsu, April 1972, p. 29; (4) figures for 1973 are based on the first section of Chapter 4 of the text.

1. The Tach'ing Refinery. Major equipment was imported
from Italy,[25] including a constant-pressure fractionating tower, a
reduced-pressure fractionating tower, and catalytic crackers. The
capacity of the Tach'ing Refinery was reportedly increased from 2.5
million tons in 1970 to 5 million tons in 1974.[26]

2. The Lanchow Refinery. The plant was constructed in the
1950s and has undergone many technical transformations. Its processing ability has been more than doubled and the variety of products rose from its original 16, to 160 in 1973.[27] By 1974 its capacity exceeded 4 million tons.

3. The Shanghai Refinery. Originally a small plant with outmoded equipment, it underwent large-scale renovations during 1960-
69. By 1969 the refinery could process ten times as much crude oil
as in 1958. Additional expansion commenced in 1970, when 4 fractionating towers, 3 heating furnaces, 16,000 meters of pipeline, and
various pumps were added to the existing facilities and again doubled
the processing capacity.[28]

4. Peking General Petrochemical Plant. This has been one
of the most significant petrochemical enterprises in China built in
recent years. The construction of the plant was begun in the winter
of 1968. Its first-stage construction, completed in 1972, includes
three major refining installations that can process 2.5 million tons
of crude oil per year. Its 15 refining and chemical installations,
when completed, will produce fuels; lubricants; benzene; paraffin;
liquid petroleum gas; and nearly 100 kinds of chemical products, including synthetic rubber, synthetic fiber, plastics, and chemical
fertilizers.[29] Capacity by 1974 had reached 4 million tons a year.

5. Dairen (Talien) Refinery. Built at the most important seaport in North China, this plant has been constantly renovated since
1960. Processing capacity rose tenfold between 1950 and 1970 and
doubled again between 1970 and 1971. By 1972 its processing capacity stood at 3.5 million tons.[30]

6. The Nanking Refinery. This refinery was also greatly expanded during the 1960s. It processes crude oil from Shengli oil
field and supplies raw materials for the growing chemical industry
in that area. Its 1974 capacity exceeded 3 million tons a year.

7. The Fushun Number 1 Refinery. Originally the chief refinery for synthetic oil in China, this plant was converted into a
crude oil refinery in the early 1970s in order to process Tach'ing
crude. Its 1974 processing capacity was 1.5 million tons a year.[31]

8. The Yumen Refinery. One of the oldest refineries in China,
the plant has undergone continued renovation. Its capacity was increased from .3 million tons in 1957 to 1.2 million in 1967 and 1.5
million in 1973.

9. The Anshan Refinery. This is a new project that was completed in January 1973, when its capacity was 2.5 million tons a year.[32]

10. The Hangchow Refinery. Another plant that has been built recently, this plant has a capacity believed to be between 1 and 2 million tons.

The revitalized and expanded refining industry significantly improved the production mix of China's petroleum products. In 1958 the country could only produce 79 varieties of petroleum products, and most of the high-grade products had to be imported. Some 60 new products were manufactured in 1962 alone, including machine oil, most of which had formerly been imported. In 1973 the Yumen Refinery was reported to be capable of making more than 120 distinctive kinds of petroleum products and the Lanchow Refinery was reported capable of producing 160 kinds. At the yearly export fair held at Canton there have been displays of petroleum products in increasing number, including heavy diesel oil, lubricants, petroleum asphalt, and paraffin wax. By 1973 there were between 350 and 400 petroleum products, as shown in Table 4.5.

TABLE 4.5

Number of Petroleum Products by Refinery, Various Years

Year	Number of Products
1949	19
1957	29-32
1958	79
1959	144
1961	175
1965	320
1973	350-400*

*It was reported that China would produce several hundred oil products in 1966. See Source 5.

Sources: (1) For 1949 and 1957, Shih-yu Kung-yeh T'ung-hsun, no. 10 (1957), p. 1; (2) for 1958 and 1959, Shih-yu Lein-chih, no. 3 p. 1; (3) for 1961 and 1965, Che'n Cheng-siang, "The Petroleum Resources of China and Their Development," in Research Report (Hong Kong: The Chinese University of Hong Kong, 1968), p. 7; (4) for 1973, from New China News Agency, Peking, March 15, 1973; (5) Chung-kuo Hsin-wen, September 1, 1966.

MAP 4.1

Crude Oil Flow in China

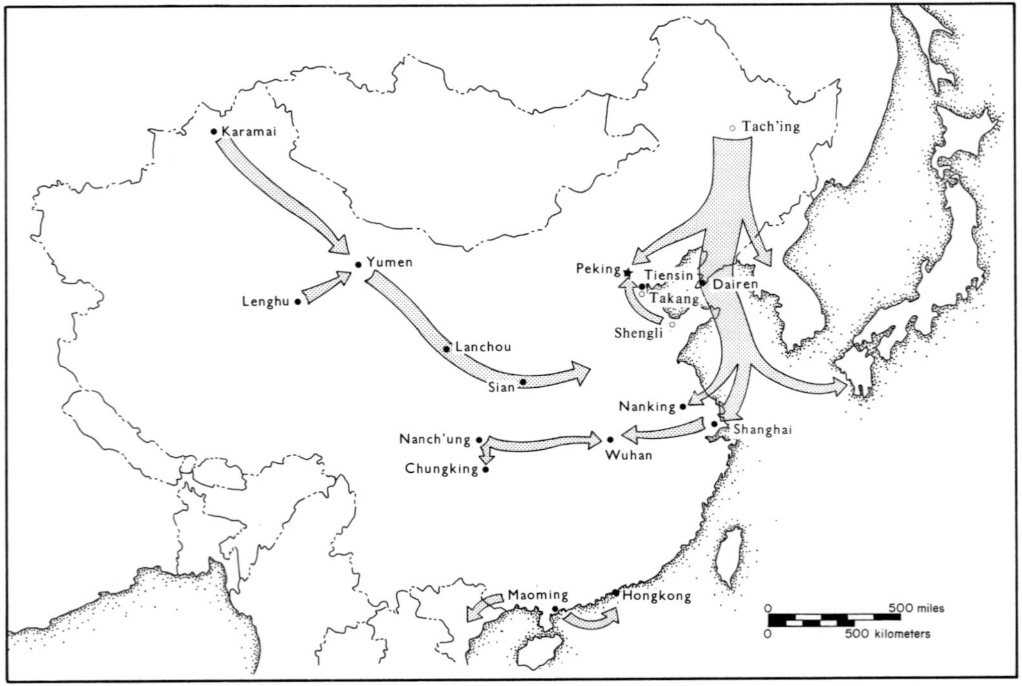

Source: Tatsu Kambara, "The Petroleum Industry in China," The China Quarterly 60 (December 1974): 713.

The location shift of the newly built refineries signifies another change in the refining industry. Until the early 1960s crude oil was produced in the northwest and in Szechwan and had to be shipped some 1,500 miles to Shanghai and Dairen for refining. The cost of transportation was obviously prohibitive. Most refineries built in the early years therefore were located near the sources of crude oil and the products were supplied to local industries. Since then, as crude oil development concentrated on the eastern coast, most of the refineries have also been built or expanded near these consuming centers. The expansion of refineries at Nanking, Shanghai, and Dairen and the construction of big refineries at Anshan, Peking, and Hangchow provide ample evidence of this trend.

Apart from balancing the distribution of oil refineries between producing and consuming centers, this shift in crude oil development also provided a direct linkage of the producing centers with the petrochemical industries in the coastal areas. Hundreds of chemical plants using petroleum products as raw materials have mushroomed around the new refineries in recent years. For example, a large number of chemical plants are now under construction at the site of the Peking Refinery, including a 1,3-cis polybutadiene plant of 3,000 tons per year, a polypropylene plant of 4,000 tons per year, an acrylonitrile plant of 3,000 tons per year, a styrene and polystyrene plant of 4,000 tons per year, a caprolactam plant of 3,000 tons per year, and a detergent plant of 1,000 tons per year.[33]

Despite real progress, the Chinese refining industry in 1975 remains weak and inadequate. It is hampered by inadequate investment; most of the net additions in refining capacity gained since the early 1960s have come from technical innovations rather than from the building of new plants. According to official reports, six old refineries doubled their processing capacity in 1970, primarily through improvements in technology. In 1971, of the net additions to refining capacity, 66 percent came from technical transformation of existing refineries.[34] The tapping of this latent capacity can increase production in the short run but cannot provide the necessary growth in the long run. Statistics in western Europe indicate that the replacement rate for European refineries is 6 percent, which would imply an average service life of 16.67 years;[35] at this rate, most of the refineries built in the 1950s and 1960s in China should be replaced by the early 1980s. On the basis of the projection in Table 4.2, China's crude oil output should reach 330 million tons by 1985. If it were further assumed that 70 percent of the crude will be processed domestically, then China would require a refining capacity of 230 million tons of crude oil a year, an increase of more than 180 million tons over the 1974 capacity. In the 1966-69 period, the capital outlay for new refineries in Western Europe was $10 per ton; on

this basis, a net increase of 180 million tons of capacity would require an investment of $1.8 billion (at the 1966 price). If these costs
are added to the cost of replacing the old refineries, China's capital
investment for the refining industry during the coming decade may
aggregate from $2 to $2.5 billion at the 1966 price. Without this
concomitant capital investment, the gap between the growth of crude
oil output and the growth of refining capacity will expand, with a
deleterious effect on the future development of the petroleum industry.

TRANSPORTATION FACILITIES

An efficient transport technology can service the vital logistic
needs of a nation's petroleum industry by gathering and distributing
the enormous quantities of oil necessary to power and lubricate the
transportation vehicles as well as to service other important production and consumption functions. Essentially, the transportation requirements of a petroleum industry fall into three categories: the
first and most significant is the shipment of crude oil from the producing wells to the refineries; the second is the various petroleum
products that are shipped from the refineries to the marketing areas;
the third, on a much smaller scale, is the products of market-
located refineries that are distributed to the final consumer.[36]

In most major oil-producing countries, crude supplies are
moved to refineries by whatever combination of pipeline and ocean
tanker transport is best adapted to the given locational and economic
factors. In general, pipelines are primarily used in producing areas,
to deliver the crude to the nearest possible export point at which
ocean tankers can take over.

This mode of transport was not utilized by China initially because of the inaccessibility of most of the oil fields to waterways as
well as because of the high costs involved in constructing and maintaining a pipeline system. Instead, railroads and highways played
the vital transport role in the development of China's petroleum industry.

Before the operation of Lanchow Refinery in 1959, most of the
crude produced in the northwest hinterland had to be shipped 2,000
miles to the east coast for refining. At that time, the only trunk line
connecting the northwest and the east was the Lung-hai railroad,
starting from Lanchow. The most difficult and costly part of this
entire transportation process was moving the crude from oil fields
to rail terminals. For example, in order to move the oil from
Tsaidam Basin in Tsinghai to the Lanchow terminus, it must be
carried by thousands of trucks for 400 to 500 kilometers, which includes negotiating a mountain pass of 3,000 feet. Truck transport

in the plateau was extremely expensive; in many cases the trucks consumed a third of their payload.[37] In Szechwan, where the transportation conditions are more favorable, the cost of transporting one ton of crude from the oil field to the local refinery was 200 yuan, compared to the production cost of 150 yuan.[38] The slow development of the Karamai and Tsaidam oil fields was mainly the consequence of a transportation bottleneck.

The first pipeline in China was constructed in 1958, connecting the oil field at Karamai with the small refinery at Tushantzu. The 174-kilometer pipeline was completed in 1959, with an annual delivery capacity of 1.1 million tons. A 60-kilometer supplementary pipeline was added later.[39] Another short-distance pipeline, of 19 kilometers, was also completed at Lenghu in Tsaidam. In the southwest area, pipeline was laid between the Nanch'ung oil field in Central Szechwan and the major industrial center of Ch'ungking.

A major pipeline was constructed in 1966, connecting the Yumen oil field with the Lanchow Refinery. This 880-kilometer pipeline greatly facilitated crude transmission in the northwest area; by the mid-1970s, however, its capacity was still rather limited.

Recent development of the pipeline system includes several important projects: (1) the construction of a 1,000-kilometer gas pipeline in Southern Szechwan, where the biggest natural-gas field is located[40] and (2) the construction of a pipeline 60 centimeters (24 inches) in diameter and 1,152 kilometers long to connect the Tach'ing oil field in Heilungkiang with the Seaport Chinwangtao on the Liaotung Peninsula.

The Tach'ing-Chinwangtao pipeline, which began its construction in August 1970, was completed in October 1973. The pipeline, with 19 pumping stations, runs through Heilungkiang, Kirin, Lioaning, and Hopei. Official reports indicate that it cuts across many hills and over 260 big and small rivers, 40 rail lines, and 200 highways. In June 1975, the Tach'ing-Chinwangtao pipeline was extended to Peking through Tientsin. The new pipeline of 355 kilometers, connects Chinwangtao and the nation's capital bringing the Tach'ing crude oil to the Peking General Petrochemical plant.[41]

Construction of a second pipeline in the northeast area, paralleling the northernmost segment of the first and terminating at T'ieh-ling in Liaoning Province, began in September 1973 and was completed by October 1974. Eventually, this line will be extended to the port of Dairen.[42] A third new pipeline of 251 kilometers connecting Shengli oil field with the port of Huangtao, near Tsingtao in Shantung Province, was also completed in September 1974.[43] Since the completion of these new projects, China in 1975 has had approximately 5,000 kilometers (3,100 miles) of pipeline. (See Table 4.6.)

MAP 4.2

Tach'ing-Chinwangtao-Peking Pipeline

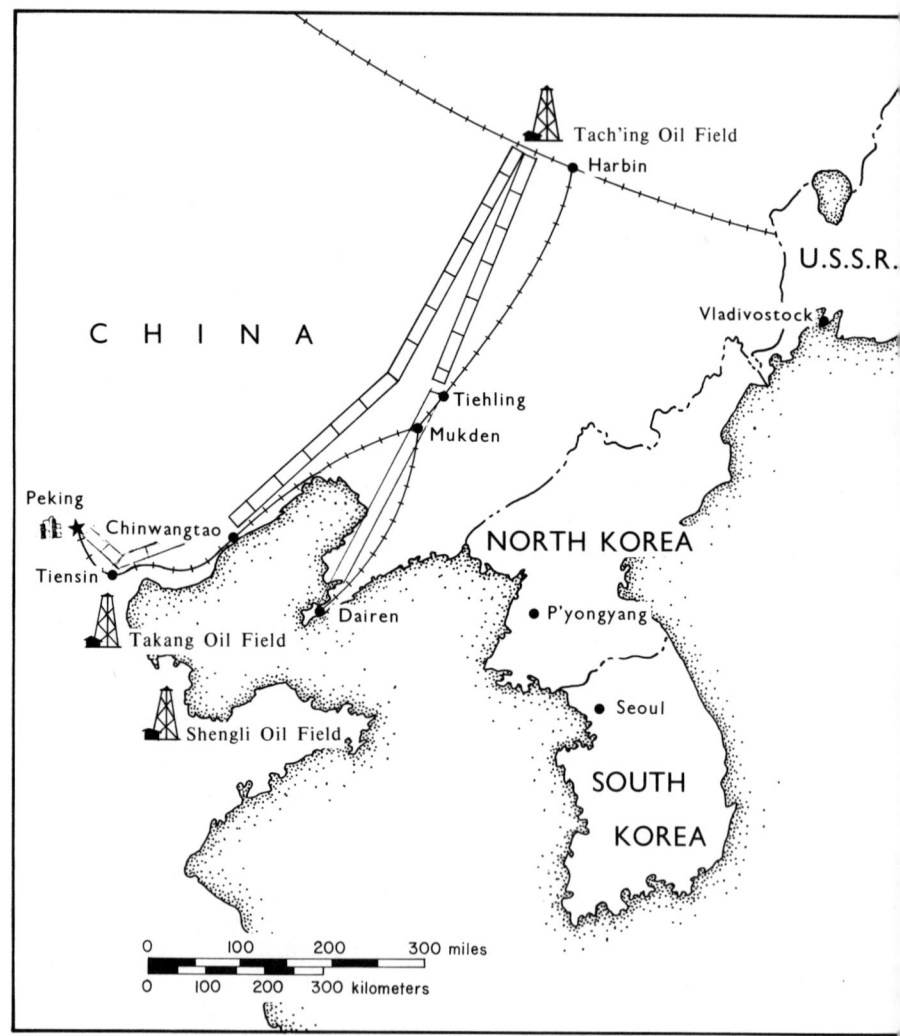

Source: Peking Review, no. 1, 1975, pp. 14-16, and no. 29, 1975, pp. 6-7.

TABLE 4. 6

Pipeline System, 1975

Pipeline	Length (in kilometers)	Diameter (in inches)	Annual Capacity (in tons)
Tach'ing-Chinwangtao	1,152	24	(12,000,000)
Chinwangtao-Peking	355	24	NA
Tach'ing-T'iehling	NA	24	NA
Shengli-Huangtao	251	NA	NA
Shengli-Peking	300	16	2,500,000
Fushun-Shenyang	NA	24	1,000,000
Lichin-Poshan	NA	14	NA
Yumen-Lanchow	880	16	NA
Karamai-Tushantzu	147	24	700,000
" "	"	16	400,000
" "	60	14	NA
Karamai-Pechientan	24	NA	NA
Tushantzu-Lanchow- Sinkiang Railway	NA	16	NA
Nanch'ung-Ch'ungking	250	16	400,000
South Szechwan Gas Pipeline	1,000	NA	NA

NA = Not available.

Sources: (1) For Tach'ing-Chinwangtao, Tach'ing-Peking Pipelines, Peking Review, no. 1, 1975, pp. 14-16, and no. 29, 1975, pp. 6-7; (2) for Tach'ing-T'iehling, Foreign Broadcast Information Service, Daily Report, December 11, 1973, p. B-1; (3) for Shengli-Huangtao, Foreign Broadcast Information Service, Daily Report, January 8, 1975, p. G-4, and Chien Yuan-heng, "An Analysis of Peking's Petroleum Production, Exportation and Transportation Capabilities," Chung-kung Yen-chiu 9, no. 4 (1975): 102; (4) for South Szechwan gas pipeline, from Peking Review, no. 4, 1973, p. 19, and Chung-kuo Hsin-wen (Chengtu), September 10, 1973, p. 2; (5) for other pipelines, from Chuan Wei, "A Critical Review on Recent Development of Petroleum Industry on Chinese Mainland," Ta-lu Ching-chi yen-chiu 7, no. 1 (January 1975): 51-52.

The construction of additional pipelines is expected to be very limited, however, because of economic and technological factors. Pipeline construction requires large quantities of seamless steel pipe, which China's steel industry cannot currently supply. Moreover, the crude oil produced in Tach'ing has a high degree of paraffin wax and a viscosity with a high solidification point. Hence, in winter, this crude must be kept at a stable temperature while being transported. The ancillary facilities needed to maintain these pipeline operations are very expensive.

There are other considerations. Broadly speaking, pipelines lack the flexibility of tankers. They are permanent fixtures, and once a large-diameter trunk pipeline system is laid it cannot be re-routed, even if conditions change and a different pattern of transport becomes desirable.[44] Therefore pipelines are generally designed for their ultimate capacity, right from the beginning. For all these reasons, the building up of a tanker fleet has received priority in recent years.

Between 1967 and 1970, nine 10,000-ton vessels were reportedly launched in Chinese shipyards. On April 2, 1969, a 15,000-ton oil tanker, the Tach'ing-27, was launched, beginning a new phase in the Chinese shipbuilding industry.[45] In 1971, four 10,000-ton freighters; a 20,000-ton tanker, the Tach'ing-28; and a 25,000-ton tanker, the Tach'ing-29, were built. Subsequently the 35,000-ton tanker Chinhu joined the tanker fleet. In 1974, Peking also purchased a 75,000-ton tanker, the Beauregard,[46] the biggest tanker China has ever had, from a Norwegian firm. In 1974 the total capacity of Chinese tankers was estimated at a miniscule 400,000 tons. As China prepares to export an increasing volume of crude oil, the pace of tanker acquisition is likely to increase sharply.

To facilitate tanker transport, major harbors along the eastern coast have been renovated, expanded, and mechanized. Investment in China's nine major ports in 1973 was reportedly more than double that of 1972 and in the first half of 1974 was three times as much as in the first half of 1973.[47] Noteworthy harbors along China's coast from north to south include Dairen (Talien), Chinwangtao, Tientsin, Tsingtao, Shanghai, Whampo, and Chankiang.

Dairen, the biggest deepwater harbor in China's northeast, can accommodate oceangoing ships alongside its wharves. A major function of this port is transporting oil; the first stage of enlarging its facilities for this purpose was completed in the middle of 1973. The port of Chinwangtao, to the west of Dairen, is northwest of the Gulf of Pohai. Here a huge oil wharf is under construction. Tientsin harbor lies at the western tip of the Gulf of Pohai and is the site of the new Takang oil field. Two berths for 25,000-ton tankers are also under construction. These three northern harbors will become the main exporting terminuses for crude oil from the Tach'ing and Takang oil fields sent to Japan and to other industrial centers in the coastal areas.[48]

Lying in the Kiaochow Bay of Shantung Peninsula, with deep calm water and a mild climate, Tsingtao harbor is both a commercial port and a shipbuilding center. The port also has under construction an oil storage tank. Shanghai, China's biggest trade port, has built or rebuilt 12 berths that can accommodate ships of up to 50,000 tons and has added over 30 percent more cargo-handling machinery since 1966. In 1973 another three berths were also rebuilt to accommodate ships of more than 50,000 tons.[49]

Chankiang harbor, which lies in the Kwanchow Bay in southern China, is the dispersal point of crude oil produced by the Maoming shale combine. It is the first Chinese port capable of handling 50,000-DWT tankers. In recent years an oil bunkering wharf has been constructed. Oil products from China that are shipped to Hong Kong and Southeast Asia must come from the Chankiang harbor.

The current expansion of the tanker fleet and harbors clearly signifies a fundamental change in Chinese transport policy from reliance on railroads to the use of pipelines and tankers, a policy that has been adopted by most oil producers in the world. Although the combined capacity of pipeline and tanker transport is still relatively insignificant, the trend is expected to gain momentum in the immediate future.

INTERNATIONAL COMPARISONS

The Chinese petroleum industry can be assessed in a more meaningful conceptual framework by the use of international comparisons.

First, in terms of petroleum reserves, the estimated 3 billion tons of probable reserves (A + B class) in China, when compared with world reserves of more than 90 billion tons, accounts for only 3 percent of the total. On the other hand, the amount of Chinese probable reserves in 1972 was approximately 54 percent of the reserves of the United States and less than one-third of those of the USSR. (See Table 4.7.)

Second, China's refinery capacity, in spite of the fact that it multiplied 37 times between 1952 and 1972, remains relatively insignificant when compared with that of other advanced countries. In January 1973, refining capacity for crude oil processing in the noncommunist countries amounted to 2.5 billion tons a year;[50] the Chinese refining capacity thus accounted for only 1.4 percent of that of the noncommunist world, or only 5 percent of the U.S. and Canadian combined capacity.

TABLE 4.7

Established World Oil Reserves and Production, 1972

| Area | Reserves | | Production | |
	Amount (in hundreds of millions of tons)	Percent of the World Total	Output (in hundreds of millions of tons)	Percent of the World Total
United States	56	6.2	5.3*	20.4
Canada	13	1.4	0.9	3.4
The Caribbean	25	2.7	1.9	7.3
Other nations in Western Hemisphere	20	2.3	0.6	2.6
Western Europe	17	1.9	0.2	0.9
Africa	39	15.3	2.8	10.8
The Middle East	485	53.3	8.9	34.2
The USSR	103	11.3	3.9	15.1
China	30	3.3	0.45	1.8
Other nations in Eastern Hemisphere	20	2.3	0.9	3.5
World total	908	100.0	26.2	100.0

*Includes 470 million tons of crude and 60 million tons of liquid natural gas.

Sources: (1) Figures for China are from Table 4.2. Sources are listed in Tables 2.4 and 2.5.; (2) the rest of the data in this table are from British Petroleum Statistical Review of the World Oil Industry (London: British Petroleum Company, 1972).

In terms of technology, the Chinese refining industry in 1975 is viewed by an American expert as comparable to U.S. or other Western refining industries of the late 1950s.[51] Existing facilities can produce a complete range of products and a variety of feedstocks for the petrochemical industry, but the quality of product is still inferior to that of those produced in advanced Western countries.

Third, China's transportation facilities present an unfavorable factor when compared with those of other advanced countries. Whereas China constructed 3,100 miles (5,000 kilometers) of pipeline during the 17 years between 1958 and 1975, the United States in the two years between 1972 and 1973 built a total of 16,770 miles. For the Free World as a whole, the newly constructed pipelines in these two years totaled 37,840 miles.[52] On January 1, 1959, the United States had about 205,000 miles of crude oil and products pipelines, including feeder lines.[53]

If China's petroleum industry is to catch up with those of other advanced countries, its capacity to explore, refine, and transport crude oil must be greatly expanded. This in turn will depend on China's capacity to make the prodigious capital investments necessary to supply the needed equipment and machinery, a topic which will be treated in subsequent chapters.

NOTES

1. Peter R. Odell, An Economic Geography of Oil (London: G. Bell and Sons, 1963), p. 8.

2. Shell International Petroleum Co., The Petroleum Handbook (London: Shell, 1959), pp. 125-26.

3. M. A. Adelman, The World Petroleum Market (Baltimore: The Johns Hopkins University Press, 1972), pp. 26-28.

4. K. C. Yeh, Communist China's Petroleum Situation (Santa Monica, Calif.: Rand Corporation, 1962), p. 28.

5. Shih-yu K'an-t'an, no. 3 (1960), p. 1.

6. Jen-min Jih-pao, July 18, 1967.

7. Shih-yu Lien-chih, no. 3 (1958), p. 11.

8. Shih-yu K'an-t'an, op. cit., p. 1.

9. Chien Yuan-heng, "Chinese Communist Petroleum Industry as Seen from the World Wide Energy Crisis," in Chung-kung Yen-chiu (Taipei), May 1974, p. 61.

10. A. A. Meyerhoff, "Development in Mainland China, 1949-68," The American Association of Petroleum Geologists Bulletin 54, no. 8 (1970): 1567-80.

11. John M. Wageman, Thomas W. C. Hilde, and K. O. Emery, "Structural Framework of East China Sea and Yellow Sea,"

The American Association of Petroleum Geologists Bulletin 54, no. 9 (1970): 1641.

12. K'o-hsueh Shih-yen [Scientific Experiment] (Peking), no. 4 (1973).

13. Christian Science Monitor, September 15, 1973.

14. New China News Agency (Peking), August 8, 1974.

15. The figure of 20 billion tons was used by Current Scene 11, no. 11 (November 1973): 21; also by Yoshio Koide, "China's Crude Oil Production," Pacific Community 5, no. 3 (April 1974): 464; the figure of 3 billion tons was suggested by Chien Yuan-heng, op. cit., p. 62.

16. Adelman, op. cit., p. 26.

17. Shih-yu Lien-chih, no. 5, 1958, p. 1, and no. 6, 1958, p. 1.

18. Brien Heenan, "China's Petroleum Industry," Far Eastern Economic Review 49, no. 13 (September 23, 1965): 567.

19. Ch'en Cheng-siang, Petroleum Resources and Their Development in China (Hong Kong: The Chinese University of Hong Kong, 1968), p. 15.

20. Shih-yu Lien-chih, no. 4 (April 1, 1958), pp. 5-9.

21. China Reconstructs, September 1971, pp. 42-44.

22. Data for Asia Research (Tokyo), no. 297 (December 5, 1961), pp. 2-10.

23. Chung-kuo Hsin-wen, September 2, 1973, p. 4.

24. Adelman, op. cit., p. 379, Appendix to Chapter VI.

25. Chien Yuan-heng, op. cit., p. 64.

26. Current Scene, September 1974, p. 20.

27. Chung-kuo Hsin-wen, September 3, 1974.

28. China Reconstructs, September 1971, pp. 42-44.

29. China Pictorial, no. 1, 1972, p. 22.

30. Chung-kuo Hsin-wen, January 12, 1973, p. 5.

31. New China News Agency (Shenyang), March 3, 1974.

32. Far Eastern Trade and Development (London), May 1973.

33. C. K. Jen, "My Impressions of the New China and Its Science and Technology," Eastern Horizon 12, no. 4 (1973): 54.

34. Chung-kuo Hsin-wen, September 3, 1973, pp. 4-5.

35. Petroleum Press Service, June 1966, p. 120.

36. Odell, op. cit., p. 139.

37. China News Analysis, March 14, 1958, p. 5.

38. Shih-yu K'an-t'an, no. 8 (1958), p. 3.

39. Ta-Kung Pao (Hong Kong), July 27, 1959.

40. Peking Review, no. 44 (November 23, 1973), p. 23.

41. Peking Review, no. 1 (1975), pp. 14-16, and no. 29 (1975), pp. 6-7.

42. Foreign Broadcast Information Service, Daily Report, December 11, 1973, p. B-1.

43. Foreign Broadcast Information Service, Daily Report, January 8, 1975, p. C-4.

44. Organization for Economic Cooperation and Development, Pipelines and Tankers (Paris: OECD, 1961), pp. 11-12.

45. Chu-yuan Cheng, "China's Industry: Advances and Dilemma," Current History, September 1971, p. 158.

46. Lloyd's List, London, April 20, 1974.

47. New China News Agency (Peking), August 10, 1974.

48. Peking Review, no. 2 (1974), p. 11.

49. New China News Agency (Shanghai), May 31, 1974.

50. The Oil and Gas Journal, January 22, 1973, pp. 39-40.

51. Bobby A. Williams, "The Chinese Petroleum Industry Growth and Prospects," Joint Economic Committee, China: A Reassessment of the Economy, July 1975, p. 245.

52. The Oil and Gas Journal, January 22, 1973, pp. 39-40.

53. Organization for Economic Cooperation and Development, op. cit., p. 21.

The existence and discovery of petroleum resources is a necessary but not a sufficient condition for the rapid development of a highly productive modern petroleum industry; sophisticated machinery and equipment are also required. In turn, this complex machinery and equipment requires a myriad of parts and an appropriate set up in order to efficiently fulfill the functions of extracting, refining, and distributing the final product.

THE ESTIMATED DEMAND FOR
PETROLEUM EQUIPMENT

The expansion of petroleum equipment manufacturing may be viewed as a function of its aggregate demand: as demand increases, so will supply. In essence, the demand for petroleum machinery and equipment arises from three major sources: fixed capital formation, repair and maintenance, and exports. The supply of petroleum machinery and equipment comes from two sources: domestic production and imports. The ratio of domestic output to aggregate demand is officially labeled the "rate of self-sufficiency."

In spite of the paucity of statistical data, one can still make a rough estimate of the aggregate demand for petroleum machinery and equipment by derivation from the capital investment data available from the First Five-Year Plan. As mentioned in Chapter 1, the total investment in the First Five-Year Plan period was given as 1.9 billion yuan,[1] of which 1,076 million yuan was for fixed capital investment.[2]

In Chinese statistics, fixed capital formation includes the purchase cost of tools and equipment as well as the value of construction

activities. Since the degree of mechanization varies significantly
for different segments of the national economy, the relative share of
capital investment in machinery and equipment purchases is also
likely to fluctuate widely. In the First Five-Year Plan period, the
relative shares of fixed investment in machinery and equipment were
officially reckoned as 40 percent for industry, 17 percent for agriculture, and 10 percent for culture and education.[3] Within the industrial sector, the relative share of capital investment for machinery
and equipment purchases was reported as 40 to 43 percent in the metallurgical industry, 46.1 percent in the First Ministry of Machine-
building Industry, and 38.4 percent in the Ministry of Electrical
Equipment.[4] With respect to the petroleum industry, no official figures are available. One source indicates that during the First Five-
Year Plan period more than 50 percent of the capital investment in
the petroleum industry was allocated for prospecting and surveying.[5]
The proportion of capital for construction, therefore, would appear
to be smaller than for other industries, whereas the share allocated
for purchasing machinery and equipment would be higher than the
three figures supplied for the metallurgical, machine-building, and
electrical equipment industries. If one arbitrarily assumes a 50
percent ratio for the petroleum industry, the capital investment for
the purchase of petroleum equipment, machinery, and tools would
approximate 538 million yuan.

   Obviously, only a portion of the aggregate demand has been
satisfied by domestic production. During the First Five-Year Plan
period, China imported 351 million old rubles in oil-well drilling
equipment, pipes, and couplings for drilling pipe from the Soviet
Union.[6] Based on the official exchange rate of one old ruble to .975
yuan, the petroleum equipment and tools imported from the Soviet
Union were valued at a total of 341 million yuan. China also purchased
petroleum machinery from Romania and East Germany, but the relative share of the Soviet Union was 96.6 percent of the total imports.[7]
Total petroleum equipment imported during the First Five-Year Plan
can be calculated at 353 million yuan (341 million yuan ÷ .966). This
imported equipment thus supported 65 percent of the aggregate demand, and the self-sufficiency rate was 35 percent.

   Two of the official statements referring to the rate of self-
sufficiency in petroleum equipment manufacturing during this period
appear to be highly contradictory. One official report claimed that
China's machine-building industry could provide only 20 percent of
the nation's needs for oil prospecting and extracting equipment in
1957,[8] while a second official report stated that while the country
could produce only 30 percent of the necessary equipment and instruments in 1952, it had become self-supporting by 1958.[9] These
contradictory statements might be partially reconciled if one assumed

that the former statement refers to key equipment for prospecting and extracting, whereas the latter could include parts and instruments. In general the 35 percent self-sufficiency rate would appear to square with the progress made in domestic production--indeed, China did not start to produce petroleum machinery and equipment until after 1955.

In the two Great Leap Forward years, 1958 and 1959, following the First Five-Year Plan, the production capacity of the petroleum industry was officially reported as having increased by 30 percent in 1958 and by 100 percent in 1959.[10] If the capacity increase was proportional to capital investment, the total investment for these two years might be estimated at approximately 1.2 billion yuan; however, investment policy during this biennium was shifted from capital-intensive to labor-intensive, with the result that the proportion of capital investment in equipment and machinery to capacity increase was probably lower than during the first plan period. If one assumes a 40 percent ratio as the ratio for all modern industry of China, then the demand for petroleum equipment and machinery during these two years would have reached 480 million yuan. In the same period, imports of petroleum equipment from the Soviet Union totaled 176.5 million old rubles,[11] or 172 million yuan. Including the equipment from other Communist countries, total imports of petroleum equipment in this biennium amounted to 177 million yuan, all of which accounts for 37 percent of the aggregate demand and yields a self-sufficiency rate of 63 percent.

In the subsequent ten years (1960-69), the production capacity for crude oil increased fivefold and refining capacity by eightfold, when compared with that of the preceding decade. The rise in capacity was not entirely the consequence of new investment; a greater portion of the increase came from technical innovations, which required little in the way of additional equipment but were primarily improvements in production methods and procedures. According to one authority, 66 percent of the increase in refinery capacity came from technical innovations in 1971.[12]

If one assumes that 60 percent of the increase in crude oil capacity and 40 percent of the increase in refining capacity stemmed from new capital investment during this decade, and if one further assumes a 3:1 proportion between crude oil and refinery investment, as in the case of the United States, Canada, and Venezuela,[13] then the capital investment flowing into the petroleum industry during this period can be estimated at 3.05 times that of the preceding decade $[(5 \times .6 \times .75) + (8 \times .4 \times .25) = 3.05]$. Since total investment in the petroleum industry has been estimated at 2.5 billion yuan for the 1950-59 time frame, it follows that investment for the 1960-69 decade approximated 7.625 million yuan ($3.05 \times 2.5 = 7.625$). It was during this time that China extensively explored the three new oil fields near the eastern coast, Tach'ing, Shengli, and Takang. A part of the equipment was transferred from Yumen as well as from other old oil fields to the new areas; for instance, more than 75.5 percent of the equipment in Yumen was transferred to other oil fields, primarily to support Tach'ing.[14] However, the investment for machinery and equipment was still relatively more important because of the large expansion of refineries. Assuming that machinery and equipment outlays constitute 45 percent of the capital investment of the petroleum industry, it follows that the investment in equipment and machinery during the 1960-69 period equaled 3,431 million yuan, 80 percent of which was domestically supplied.

During the 1970-74 period, when China doubled her crude oil production, from 30 to 63 million tons, investment in refining, transportation, and production expanded rapidly. Authenticated purchases of petroleum equipment from abroad during these five years totaled to 500 million dollars, the equivalent of 1.2 billion yuan.* Assuming the officially announced self-sufficiency rate of 80 percent to be correct, the demand for petroleum equipment can be estimated at 6 billion yuan. Since by the early 1970s the Chinese petroleum industry had entered a stage of advanced mechanization in production, the requirements for machinery and equipment occupied a much higher proportion of capital investment than heretofore. Thus if one assumes 55 percent of the capital investment to be in the form of machinery and equipment, the capital investment in the petroleum industry should be 10,900 million yuan for 1970-74, an amount almost equal to the total invested during the 1953-69 period. (See Table 5.1.)

The total capital investment in the petroleum industry in the 22 years between 1953 and 1974 can thus be estimated at 21 billion yuan, or approximately $9 billion, of which 10.5 billion yuan, or $4.5 billion, were allocated for petroleum equipment and machinery. Since the export of petroleum equipment from China was negligible, this demand can be considered as equal to the aggregate demand for petroleum equipment. One-fourth of this aggregate demand, or $1 billion, was imported from abroad, and the remainder was supplied by domestic production.

THE PROGRESS OF DOMESTIC PRODUCTION

Prior to 1949, with the exception of a few factories making maintenance equipment and parts, there were virtually no specialized

---

*This figure represents realized imports, not contracts signed.

TABLE 5.1

Estimated Fixed Capital Investment in the Petroleum Industry
and Aggregate Demand for Petroleum Equipment, 1953–74

Period	Fixed Capital Investment in Petroleum Industry (in millions of yuan)	Petroleum Equipment			
		Aggregate Demand (in millions of yuan)	Domestic Supply (in millions of yuan)	Imports (in millions of yuan)	Self-Sufficiency Rate (in percent)
1953–57	1,076	538	185	353	34
1958–59	1,200	480	303	177	63
1960–69	7,625	3,431	2,745	686	80
1970–74	10,900	6,000	4,800	1,200	80
Total for 1953–74	20,801	10,449	8,033	2,416	77

Source: Derived by the author in the first section of Chapter 5 of the text.

plants in the field of petroleum equipment in China. Between 1949 and 1952, most of the needed equipment was supplied by the Soviet Union. As the country started extensive geological surveying and prospecting during the First Five-Year Plan period, the demand for drilling machines and tools increased rapidly. In 1954 a specialized task force was authorized to convert a number of small-scale plants in Shanghai that had originally produced printing machines and parts for textile machinery, to producers of basic oil-drilling tools and parts. With the opening of new oil fields in the Karamai and Tsaidam Basins in 1955, these plants were rapidly expanded. Because of the primitive technology employed, however, their production was confined to relatively unsophisticated oil extracting machines, drilling tools, oil pumps, and small compressors.

To supplement these inadequate equipment-producing facilities, two major projects commenced construction in Lanchow during 1956, the Lanchow Petroleum Machinery plant and the Lanchow Petroleum and Chemical Machinery plant. They were both designed to produce a new integrated series of modern petroleum drilling equipment.[15] The facilities, which were completed in 1960, were financed by Soviet aid and initially produced oil equipment.

During this period, trial runs began at several specialized plants in Sian, Paochi, and Shanghai that manufactured prospecting instruments, drilling rigs capable of drilling to 1,000-1,200 meters, various types of drill bits, and a series of instruments for the control of drilling operations. However, the technology employed was too primitive and the product quality unsatisfactory.

With the Great Leap Forward in 1958, more than two dozen specialized plants and a few multiple-product factories were established in Shanghai, Harbin, and Tientsin. These plants produced large quantities of drilling rigs with 1,000-1,200 meter capacity. Experimental prototypes were built, and trial runs at producing tower facilities for small refineries were conducted. Despite a marked increase in capital investment during the 1958-60 period, annual imports of petroleum equipment from the Soviet Union remained nearly constant, indicating that domestic output produced an increasing percentage of the new equipment used in Chinese oil fields.

The opening of Tach'ing in 1960 provided a new impetus for petroleum equipment manufacturing. Until 1960 the petroleum and chemical equipment industry was only a subsection of the heavy mining machine industry, but since then it has become a separate branch of China's machine-building industry. Output of petroleum equipment in 1963 was officially reported as 60 percent higher than that in 1962, and by 1964 the output had doubled again.[16] Since 1963 the central planners have placed a high priority on the development of refining equipment. A dozen heavy machinery plants in Shanghai, Shenyang,

and Harbin were converted into refining equipment producers. The
first fractional distillation tower was successfully built by the
Hsin-chien machinery plant in Shanghai in 1964, and a wing-type
heat exchanger, another key piece of refining equipment, was produced at the Shanghai Szu-fang boiler plant in the same year. High
pressure columns were also manufactured on a trial basis at this
time. By 1964 the industry was capable of producing coke towers
for decomposing heavy oil into light petroleum, main-column extracting installations for making lubricating oil, heat exchangers,
and various oil pumps. A large vacuum fractionating tower 27 meters
high, 290 tons in weight, and 6.4 meters in diameter was produced
in Shanghai during this period. With these developments, the petroleum equipment industry began to acquire meaningful proportions.[17]

Progress has been continuous since 1965, and the available
evidence indicates that the petroleum equipment section has tended
to outperform the machine-building industry as a whole. For instance, in terms of value of output, Chinese petroleum equipment
production has experienced a much faster growth rate than that which
characterizes her other machine industries. Between 1957 and 1971
the output value of China's entire machine-building industry increased
twelvefold, and the value of the petroleum equipment output was
listed among the above-average growth sections.[18] In 1971, when
the total output value of the machine-building industry increased 15.3
percent in Liaoning, output of petroleum equipment rose 30 percent.[19]
Understandably, petroleum equipment now ranks as one of 13 major
sections of China's machine-building industry.[20]

Although it has been unable to produce offshore drilling equipment, some petrochemical equipment, and a few of the more sophisticated instruments, the Chinese petroleum equipment industry is
now capable of producing a wide range of products that can satisfy
nearly 80 percent of the country's prospecting and exploration needs.
It now produces different types of geophysical instruments, including
magnetometers, airborne magnetometers and electromagnetic instruments, and instruments for radio and seismic surveying. The
exploration equipment produced in China includes deep-well drilling
rigs capable of penetrating 3,200 meters; pressure tools capable of
producing 500-atmosphere pressures for crushing rock layers and
increasing oil extractions; drilling tools (bits, collars, and pipes);
well-logging machines; Christmas trees; centrifugal pumps; blow-out
preventers; and gas and water separators. In the field of refining,
China now manufactures entire refining installations for a designed
capacity up to 2.5 million tons, such as the newly constructed Peking
General Petrochemical plant. In the field of transport and storage,
China now builds ocean and coastal tankers with up to 25,000 tons
displacement, steel pipes with diameters up to 30 inches, and various
sizes and types of storage containers.

In terms of technology, the Chinese petroleum equipment industry initially imitated the Soviet designs in the 1950s, but has tended to move toward independent designs in the 1960s. During the past decade it has demonstrated its capability of modifying foreign equipment in order to adapt it to Chinese conditions; the new products have generally tended to be more compact in size, lighter in weight, simpler in structure, and easier to operate than their prototypes. One example is a drilling rig produced by the Lanchow petrochemical equipment plant. In the past the major product of this plant was a high-powered rig that had been copied from Soviet designs and was said to be complicated, clumsy, heavy, inefficient, and dangerous to operate. In 1970, however, the Chinese redesigned the rig and successfully produced a new model with 60 percent more power that was one-fifth lighter. [21] Another example is the transformation of the Lanchow Oil Refinery, a plant also designed and constructed with Soviet aid. Originally the plant was designed to refine one type of crude oil and produce 16 kinds of petroleum products, but after design modifications by the Chinese in 1962, the plant could be utilized to refine three types of crude oil (that is, paraffin-base crude oil, asphaltic base crude oil, and mixed base crude oil) and produce 160 kinds of petroleum products. [22] This capability for making design improvements has helped the Chinese to achieve productivity increases in the petroleum industry with relatively small additions in equipment.

MAJOR PRODUCERS AND PRODUCTS

After 20 years of development, a network of petroleum equipment manufacturing has been established in China. Presently there are more than 100 specialized plants in the fields of oil prospecting and control instruments, drilling machinery, tools, refining equipment, oil tankers, oil pipes, and equipment for the petrochemical industry. On the basis of diverse sources, I have outlined the major producers and their principal products.

The Shanghai Area

Of the 100 petroleum equipment producers, 40 are located in Shanghai, the leading industrial center in China. Half of these producers specialize in drilling machines, derricks, valves, bits, and spare parts, while the remainder produce boilers, air compressors, and oil-refining equipment.

Shanghai Hsin-chien
Machinery Plant

The Shanghai Hsin-chien machinery plant was originally a
small shipyard and during the 1960s was renovated in order to manufacture refining equipment. It successfully turned out the first
fractional distillation tower in 1963. After initial testing, the new
equipment met the requirements for sustaining high temperatures
and pressures while remaining resistant to knocking and fire.[23] It
is now the major plant for producing petroleum refining installations.

Shanghai Ta-lung Machinery Plant

The Shanghai Ta-lung Machinery Plant was an old plant established in 1902 for producing farm tools and textile machines that was
transformed into a producer of petrochemical equipment. Its main
products include annealing and pressurizing equipment, including a
600-ton nonferrous metal pressurizing machine that is as high as a
three-story building and has a five-ton heat annealing manipulator.[24]
In recent years the plant has turned out large quantities of chemical
fertilizer equipment, such as a 3,300 cubic-meter-per-hour carbon
dioxide compressor for use in the production of nitrogen fertilizer.
Also included in its line of products are large freezers making
25,000 tons of synthetic ammonia a year, high-pressure circuit
pressurizers, high-pressure copper liquid pumps, ammonia liquid
pumps, and carbon dioxide pressurizers that can produce 40,000
tons of urea a year. In 1965 it employed more than 2,000 workers.[25]

Shanghai Szu-fang Boiler Plant

The Shanghai Szu-fang Boiler Plant has produced oil-refining
equipment for the wing-type heat exchanger since 1964. The quality
of its products is comparable to that achieved by the technologically
advanced nations of the world.[26]

Shanghai Boiler Plant

One of the major manufacturers of power equipment in China,
the Shanghai Boiler Plant produced the first large vacuum fractionating tower in 1964. This tower is 27 meters in height, 90 tons in
weight, and has a diameter of 6.4 meters.[27] The firm has since
become the principal supplier of fractionating towers.

Shanghai Petroleum Machine
Parts Plant

Transformed from a textile machine repairer into a producer
of petroleum drilling tools in 1956, the Shanghai Petroleum Machine
Parts Plant produced four categories of a total of 52 varieties of
drilling tools by 1961. It also produces 300 meter treble-toothed
bits as well as bits for soft and very tough strata; that is, the three-
cone hard-formation bits and soft-formation drills.[28]

Shanghai Li-shen Machinery Plant

The Shanghai Li-shen Machinery Plant is a general machinery
plant that was renovated into a petroleum equipment producer in 1958.
The major product of this firm is C-1000 meter drilling rigs for
shallow- and medium-depth oil wells. Production in 1964 was reported as 100 light drilling rigs per year.[29]

Shanghai Yung-feng Machinery Plant

The Shanghai Yung-feng Machinery Plant specializes in producing oil valves and control equipment. Its main products are the
Christmas-tree type machine, with a working pressure of 200 pounds
per square inch.[30]

Shanghai Shaped Steel Tubing Plant

The Shanghai Shaped Steel Tubing Plant was set up in 1953 and
started to produce seamless steel tubes for petroleum equipment in
1958. Between 1958 and 1970, 2,100 varieties of ferrous and nonferrous seamless pipes, shaped steel tubes, and high-grade alloy
steel materials were produced by this plant.[31]

Shanghai Petro-Chemical
Machinery Plant

The newly established Shanghai Petro-Chemical Machinery
Plant specializes in producing distillation columns used for separat-
ing the reformed oil into gas and motor spirits.

Shanghai Deep-Well Pump Factory

Before 1970, the Deep-Well Pump Factory mainly produced
deep-well pumps copied from Soviet models, but it has since

manufactured a deep-well submerged pump of its own design. Annual output as of 1971 was reported to be 2,000 units.[32]

The Lanchow-Paochi Area

In northwest China, where several major oil fields are located, manufacturing of petroleum equipment was developed rapidly in several cities surrounding Lanchow, which is now second only to Shanghai as a major producing center.

Lanchow Petrochemical Machinery Plant

Built with Soviet aid in the 1950s, the Lanchow Petrochemical Machinery Plant in 1960 produced oil drilling rigs that were 35 meters high, comprised of 21,000 parts, and capable of drilling to a depth of over 3,000 meters.[33] In 1970 the plant redesigned a Soviet rig and produced a new type of power drilling rig that is considered a major technological breakthrough.[34] The plant is now China's chief producer of medium-depth well-drilling machines.

Lanchow Petroleum Machinery Plant

The Lanchow Petroleum Machinery Plant was also built with Soviet aid in the 1950s. It was designed to produce an integrated series of modern petroleum drilling equipment.[35] However, very little information about this plant's production is available for the years after 1956.

Lanchow General Machinery Plant

Formerly an old artillery weapons factory built half a century ago, the Lanchow General Machinery Plant has been converted to producing an integrated system of drilling equipment. In 1973 the plant employed more than 3,000 workers and produced a complete oil extraction unit totaling 7,000 tons, including subsurface centrifugal pumps as well as subsurface plunger pumps.[36] Its products were displayed in the 1973 Spring Trade Fair at Canton and were praised by the Chinese press as being quality-competitive with international standards.[37]

The First Machinery Plant of Paochi

The First Machinery Plant of Paochi has been a major supplier of derricks and drilling tools since the 1950s. Its principal products are light drill rigs, bits, rotary swivels, drill pipe, drill collars, and tool joints.[38] In 1958 the plant was selected as an advanced producer of petroequipment.

Paochi Petroleum Pipe Plant

Newly built, the Paochi Petroleum Pipe Plant specializes in producing steel pipes for transporting crude oil and natural gas. In the past it was able to manufacture only pipe with a diameter of 16 inches, but in 1973 began to produce pipe with a diameter of 30 inches.[39]

Tienshui Oil Pump and Oil Valve Plant

Built 150 miles southeast of Lanchow in 1973, the Tienshui Oil Pump and Oil Valve Plant specializes in producing various kinds of oil pumps and oil valves.[40]

Yumen Machinery Plant

Originally the repair workshop for the Yumen oil field, the Yumen Machinery Plant was converted into a petroleum equipment plant in the late 1950s. In 1958 the plant turned out rigs capable of drilling 600 meters, tooth bits, and tube-type deep-well pumps.[41] In 1959 it began to produce drill joints and mud-drilling pumps. By 1964 it could produce ten different types of deep-well pumps.[42]

Sian Geophysical Instruments Plant

The Sian Geophysical Instruments Plant is a major supplier of instruments for geophysical exploration, and its products include gas-logging instruments, autoelectrical logging instruments, radiometers, apparatus for determining velocity by ultrasonic waves, autoelectrical instruments for field work, high-speed thermal logging instruments, apparatus for determining the resistance of liquid in bore holes, low frequency seismodetectors, and low frequency seismographs, as well as M-I multiple pumps, magnetometers, 203-type airborne magnetometers, 203-type isotope well-testing instruments, and geophones.[43]

The Peking-Tientsin Area

With the opening of new oil fields in Takang in northern China, petroleum equipment plants were also set up in Tientsin, Peking, Taiyuan, Hantan, and Changchiakou.

Tientsin Chemical Engineering and
Petroleum Equipment Plant

The Tientsin Chemical Engineering and Petroleum Equipment
Plant specializes in the manufacturing of high-pressure pumps and
chemical equipment such as urea synthesis towers and carbon dioxide
compressors.

Shihchiachuang Coal-mining
Machinery Plant

In 1972 the Shihchiachuang Coal-mining Machinery Plant turned
out a new type of shallow-stratum petrol drilling rig that is light,
compact, and easier to operate and maintain than existing drilling
rigs in the same category. Its efficiency is said to be two or three
times higher than that of existing rigs. The new rig is now being
mass-produced. [44]

Changchiakou Mining Machinery
Plants Number 1 and Number 2

The Changchiakou Number 1 Mining Machinery Plant produced
light drilling rigs capable of drilling 1,000 meters, shallow-well
pumps, drill bits, and 1,000-meter autodrilling rigs in 1961-64, while
the Number 2 plant turned out 80 M^3 and 40 M^3 drilling mud pumps
and 20 M^3 oil tanks. In 1970-71 these two plants cooperated in producing 30-meter-high derricks, 2,000-meter drilling rigs, and
3,000-meter mobile drilling rigs. [45]

Liberation Army 3603 Plant
in Peking

The Liberation Army 3603 Plant in Peking makes containers
for storing and transporting petroleum. Its main product is the
horizontal metal oil drum. In 1972 a new type of oil drum of high
quality was produced by this plant. [46]

Taiyuan Mining Machinery Plant

The Taiyuan Mining Machinery Plant turned out China's first
large oil-well drilling machine capable of boring 1,200 meters, in
1957. The machine is modeled on Soviet design and weighs 65 tons. [47]
The plant now specializes in oil well equipment for medium-depth
drilling.

Manchuria Area

In the northeast region, Harbin and Shenyang are the two major centers for producing petroleum equipment. With the development of the Tach'ing oil field in this region, the demand for locally supplied petroleum equipment and parts has grown significantly.

Harbin Boiler Plant

Built with Soviet aid in the 1950s, the Harbin Boiler Plant is one of the country's best-equipped plants for manufacturing medium- and high-pressure boilers. In 1964 it successfully produced the K. L. asphalt separation tower used in the alkylation process.[48]

Shenyang Air Compressor Plant

The Shenyang Air Compressor Plant is one of China's major makers of large air compressors, which are key equipment for high-pressure refining. In 1966 it produced the first symmetrical air compressor for urea synthesis, which is capable of producing 20,000-40,000 tons of urea per year. The equipment produced by this plant was praised by the Chinese authorities as meeting international standards.[49] Together with the Ta-lung machinery plant in Shanghai, it provides most of the domestic equipment for urea synthesis.

Shenyang Water Pump Factory

The Shenyang Water Pump Factory is one of China's largest pump producers. Its products are supplied to 132 factories in 20 regions of the country. The plant began manufacturing oil-cracking pumps, which are important equipment for oil refineries, in 1961. The high-temperature- and erosion-resistant pump sends crude oil to the cracking stills for refining under high pressure. The pump has boon in mass production since 1962.[50]

Anshan Seamless Steel Pipe Plant

A component plant of the Anshan iron and steel complex, the Anshan Seamless Steel Pipe Plant is the country's chief supplier of seamless steel pipes. Built with Soviet aid in the 1950s, its current capacity is 900,000 tons a year.

Dairen Hungch'i Shipyard

The Dairen Hungch'i Shipyard is a major shipyard specializing in oil tanker building. It built the first 15,000-ton tanker, the Tach'ing-27, in 1969; another 15,000-ton tanker, the Tach'ing-29, in 1970; a 20,000-ton tanker, the Tach'ing-30, in 1971; and a 25,000-ton tanker, the Tach'ing-32, in 1972. Its capacity in 1971 was 2.7 times what it had been in 1965.

Wuhan-Canton Area

Wuhan and Canton are the two machine-building centers that have established plants for petroleum equipment in central and southern China.

Wuhan General Machinery Plant

The Wuhan General Machinery Plant began to produce petroleum equipment in the late 1960s. In 1970 it turned out a hydraulic rotary drilling rig capable of boring 600 meters.[51]

Kwangchow Heavy-Duty Machinery Plant

The Kwangchow Heavy-Duty Machinery Plant is the chief producer of chemical and petroleum equipment in southern China. Its major products include extract evaporators, heaters, and extraction towers.

GENERAL APPRAISAL

The available information on these major producers, although fragmentary and incomplete, provides a basis for assessing the strengths and weaknesses of the Chinese petroleum equipment industry.

This industry has experienced a continuous growth in quantity and quality of products. Its ability to provide equipment for the rapidly growing crude output has been augmented by the progress of China's metallurgical industry. In the 1950s, China produced only a meager quantity of annealed steel for the construction of high-pressure containers. Most alloy steel, seamless steel pipes, and stainless steel plates had to be imported. After 1964 China began to produce low-alloy steel, which is superior in tensile strength and resistant to erosion and abrasion. It is also easier to cut, form, and weld, and possesses a durability 30-100 percent greater than carbon steel. China produced only 14 types of low alloy steel in 1966, but more than 80 by 1971.

Moreover, since 1969 the Chinese metallurgical industry has been producing low-temperature-resistant steel plates for use in the manufacture of oceangoing vessels; and composite stainless steel plates for use in the manufacture of nitrogen fertilizer equipment; and high-pressure, hydrogen-resistant steel pipes for use in the manufacture of pipes for petroleum.

In 1971 China's metallurgical industry trial-produced nearly 100 kinds of new steel, many of them vital to petroleum equipment manufacturing, among which were alloy steels to make large-diameter thin-wall pipe and hard-melting alloy steels with 2,000-3,000 degree melting points.[52] In 1972 many new steel products for petrochemical equipment were added to the production list, including the 630 millimeter revolving welding tubes used in the petroleum industry, the alloy steel tubing used in fission chemistry, special steel tubing that resists temperatures as low as -120 degrees, and high temperature resistant magnesium alloy plate.[53] The increase in the number of steel products suitable for petroleum equipment will undoubtedly enhance China's ability to meet her growing demand.

There are, however, several factors that hamper the Chinese effort to achieve complete self-sufficiency in petroleum equipment.

First, as shown in the preceding section, most of the petroleum equipment producers are old plants that were originally engaged in other endeavors. Only the two plants built in the 1950s at Lanchow with Soviet aid were designed for petroleum equipment manufacturing. Not only are the production facilities of most of the plants antiquated and marginal for petroequipment building, but there is also a lack of standardization in their products. As a consequence, parts are not interchangeable among the plants and there are serious deficiencies in repair and maintenance.[54]

Second, as is frequently the case in Soviet enterprises, the most important success indicator for a plant is the fulfillment of production quotas. The preoccupation of the factory manager is to fulfill or overfulfill the assigned quota on output value or volume, and frequently these goals have been attained at the sacrifice of quality and variety. Numerous reports in the official Chinese presses indicate that although many plants meet their quantity quotas, few of them meet the requirements for variety of types.[55] In 1971 official campaigns were launched criticizing the myopic attitude associated with simply fulfilling the quantity planned at the expense of the quality and variety specified in the contracts. It was further disclosed that most plants attached so much importance to the production of the petromachine assigned that they neglected to produce the ancillary spare parts.[56] These problems continue to plague the petroleum equipment industry and impinge upon its ability to service the petroleum industry.

Third, the Chinese machine-building industry still lacks the capacity to provide large-size heavy-duty machine tools that can be used for processing other machines. The large hydraulic presses produced in China today are 12,000-metric-ton presses first manufactured in 1965, compared with the 68,000-metric-ton presses produced in the United States since 1956. Without heavy-duty machine tools, much of the large-sized petroequipment has had to be fabricated in separate parts and subsequently forged together. This has not only affected its durability and efficiency, but has also significantly increased the cost and the time required to manufacture it.

For these and other reasons, the petroleum equipment produced in China today still lags behind that of the advanced industrial countries both in quality of technology and in variety of configurations. For instance, China can only produce drilling rigs for shallow and medium-depth wells, up to 10,000 feet, or 3,200 meters; heavy rigs capable of drilling to 20,000 feet, or 6,400 meters, are not being produced in China.

In the area of refining equipment, China can now produce high-pressure cylinders up to 1,500-2,000 atmospheres for compressors and reactors, whereas in the United States 5,000-7,000-atmosphere compressors and pumps are very common.

Similar deficiencies manifest themselves in the transportation equipment sector. The largest crude-oil pipe produced in China today has a diameter of only 30 inches, although 48-inch oil pipes were being widely used in the West.[57]

China's capability to build oil tankers has advanced rapidly in recent years, but the largest tanker built in China today is still in the 25,000-ton class. Compared with the very large crude carrier (VLCC) of 250,000 to 400,000 tons produced by Japan, China's technology and capacity is still 20 or 30 years behind.

From these comparisons we can hypothesize that for China to attain the levels of technology and production of the Western countries and Japan may well require another two decades. During this period, China may still rely on foreign supplies for offshore exploration equipment, deep-well drilling equipment, petrochemical plants, large-diameter oil pipes, and large-sized tankers. It appears quite probable that China may continue to import 20 percent of her petroleum equipment to complement her domestic supply, until the mid-1980s.

NOTES

1. Jen-min Shou-tse 1958 [People's Handbook] (Peking: Ta Kung Pao, 1958), p. 473.

2. Chu-yuan Cheng, China's Allocation of Fixed Capital Investment, 1952-1957 (Ann Arbor: Center for Chinese Studies, The University of Michigan, 1974), p. 55, Table 22.

3. Chi-hsieh Kung-yeh, no. 15 (1955), p. 8.

4. Ts'ai-cheng, no. 2 (1956), p. 29.

5. Wang Hsin-san, Ti-i-ko Wu-nien-chi-hua chung ti Jang-liao Kung-yeh (Peking: Chung-hua Chung-kuo K'o-shueh Chi-shu Pu-shi Hsie-hui, 1956), p. 27.

6. Estimated based on figures provided by K. C. Yeh, Communist China's Petroleum Situation (Santa Monica, Calif.: Rand Corporation, 1962), p. 23.

7. Chu-yuan Cheng, The Machine-building Industry in Communist China (Chicago: Aldine-Atherton, 1971), p. 172.

8. Jen-min Shou-tse, 1958, p. 469.

9. Shih-yu Lien-chih, no. 3 (1958), p. 12.

10. Shih-yu K'an-t'an, no. 5 (1960), p. 4.

11. Yeh, op. cit., p. 23.

12. Ibid.

13. This ratio is estimated based on data from Chase Manhattan Bank, Capital Investments of the World Petroleum Industry (New York: the Bank, 1971).

14. China Pictorial, no. 2 (1974), pp. 5-6.

15. New China News Agency (Lanchow), October 6, 1956.

16. Chi-hsieh Kung-yeh, no. 20 (1964), pp. 14-15.

17. Cheng, The Machine-Building Industry in Communist China, op. cit., p. 305.

18. Li Feng, "Rapid Growth of Our Country's Machine-building Industry," in Wo-men Cheng Tsai, Ch'ien-chin (Peking: Jen-min Ch'u-pan-she, 1972), p. 55.

19. New China News Agency (Shenyang), January 10, 1972.

20. Chu-yuan Cheng, The Machine-building Industry in Communist China, op. cit., pp. 140-41.

21. New China News Agency (Lanchow), April 16, 1970.

22. Kung-jen Jih-pao, March 25, 1966, p. 2.

23. Chieh-fang Jih-pao, October 11, 1964, p. 1.

24. Chi-hsieh Kung-yeh, no. 20 (October 25, 1964), pp. 18-20.

25. New China News Agency (Peking), March 17, 1965.

26. Chieh-fang Jih-pao, October 11, 1964, p. 1.

27. New China News Agency (Peking), July 18, 1964; also China Reconstructs, May 1964, p. 28.

28. New China News Agency (Shanghai), April 18, 1961.

29. Chieh-fang Jih-pao, October 11, 1964, p. 1.

30. Chi-hsieh Chih-tso, no. 9 (1959), p. 8.

31. New China News Agency (Shanghai), July 25, 1971.

32. Jen-min Jih-pao, October 9, 1971.

33. New China News Agency (Lanchow), April 14, 1960.

34. New China News Agency (Lanchow), April 16, 1970.

35. New China News Agency (Lanchow), October 6, 1956.

36. Chung-kuo Hsin-wen, August 14, 1973, p. 2.

37. Ching-chi Tao-pao, April 25, 1973, p. 15.

38. Shih-yu K'an-t'an, no. 7 (1959), p. 3.

39. Chung-kuo Hsin-wen, October 26, 1973.

40. Wen-hui-pao, March 1, 1974, p. 2.

41. Shih-yu K'an-t'an, no. 7 (1959), pp. 2-3.

42. Chung-kuo Hsin-wen, April 22, 1964, p. 5.

43. New China News Agency (Sian), June 1, 1958.

44. New China News Agency (Peking), March 28, 1972.

45. Ho-pei Jih-pao [Hopei Daily] (Shichiachuan), September 23, 1971; also New China News Agency (Peking), December 7, 1971.

46. K'o-hsueh Shih-yen [Scientific Experiment] (Peking), no. 5, May 1973, pp. 16-17.

47. New China News Agency (Peking), December 12, 1957.

48. Chi-hsieh Kung-yeh, no. 14 (July 1964).

49. Jen-min Jih-pao, January 24, 1966.

50. Kung-jen Jih-pao, February 1, 1964.

51. New China News Agency (Wuhan), August 3, 1970.

52. Special Issue for China Export Commodities Fair Autumn 1972 (Hong Kong: Ching-chi Tao-pao), October 20, 1972, pp. 20-23.

53. Ching-chi Tao-pao, April 25, 1973, p. 15.

54. Shih-yu K'an-t'an, no. 1 (1960).

55. Jen-min Jih-pao, April 11, 1960, p. 24.

56. Jen-min Jih-pao, November 26, 1971.

57. Shell International Petroleum Company, The Petroleum Handbook (London: the Company, 1959), p. 325.

6

THE IMPORTATION OF
PETROLEUM EQUIPMENT

As noted in the preceding chapter, by the mid-1970s China's petroleum equipment industry could satisfy approximately 80 percent of her domestic needs. During the early years of development, China relied heavily on the Soviet Union for supplies of petroleum equipment and machinery, but after 1961, as Sino-Soviet relations deteriorated, China began to look toward Japan and the West for equipment and technical know-how, and her purchases of advanced equipment for offshore drilling, refining, and petrochemical production have increased steadily. The purpose of this chapter is to survey the changing pattern of China's overseas procurement of petroleum equipment and technology and to assess its future trends.

PETROLEUM EQUIPMENT IMPORTED
FROM THE SOVIET UNION

When the People's Republic of China was formally founded in October 1949, the Soviet Union was the first country to establish diplomatic relations with her. At that time the Chinese leaders looked upon the USSR as the most reliable potential source of the wherewithal for economic rehabilitation and reconstruction. In his famous speech "On People's Democratic Dictatorship," delivered to the sixth session of Seventh Central Committee of the Chinese Communist Party in June 1949, Mao Tse-tung clearly ruled out the possibility of British or American aid and emphasized that "internationally, we belong to the side of the anti-imperialist front headed by the Soviet Union and so we turn only to this side for genuine and friendly help, not to the side of the imperialist front."[1]

Great expectations of economic assistance from the USSR led to the formation of Mao's "lean-to-one side" foreign policy, which linked China to the Soviet Union during her first decade of growth. Immediately following the establishment of the new government, Mao led a delegation to Moscow to conclude a series of treaties and agreements that had a strong effect on the economic development of the People's Republic of China.

During China's First Five-Year Plan period, the Soviet Union aided China in the building of 141 industrial enterprises as well as in equipping other projects with Soviet-made machinery. The Soviet Union also provided China with more than 21,000 sets of scientific and technical documents. Thousands of Chinese specialists and workers were trained in the Soviet Union, while more than 10,000 Soviet experts and technicians were dispatched to China.[2]

The number of projects backed by Soviet aid and know-how gradually expanded throughout the initial four years of the First Five-Year Plan, and by 1959 the number of Soviet-aid projects had reached 291. Many of these projects, however, did not materialize because of the Sino-Soviet split in 1960. Of the Soviet-aided projects, 156 were constructed during the 1949-59 period, including giant iron and steel complexes; nonferrous metallurgical enterprises; coal mines; oil refineries; chemical plants; power stations; and factories for the production of heavy machinery, precision instruments, automobiles, airplanes, and tractors. Together they constituted the backbone of China's industrial program. (See Table 6.1.)

Although only two of the 156 projects were directly related to the petroleum industry and an additional two to the manufacture of petroleum equipment, Soviet aid gave a definite impetus to the development of the Chinese petroleum industry during this decade. As Wang Chih-ch'un, director of the General Office of the Ministry of Petroleum Industry commented, "All the aid (from the Soviet Union) played an important part in accelerating the development of our petroleum industry."[3] In terms of new capacity in crude oil production, Soviet-aided oil projects accounted for 51.4 percent of the newly installed capacity during 1953-57.[4]

Since the Soviet Union was China's main source of foreign aid, most of the oil equipment imported during the 1950-60 period came from the Soviet Union. A Russian source offers the following estimate of the ratio of the machinery imported by China from the USSR to the aggregate of all Chinese machinery imports for the years 1949-59 as follows:[5]

Type of Machinery	Ratio (in percent)	Type of Machinery	Ratio (in percent)
Machine tools	77.9	Instruments	74.7
Diesel engines	91.4	Petroleum equipment	96.6
Boring machines	84.5	Ferrous-metal machinery	57.0

TABLE 6.1

Distribution of 156 Soviet-aided Projects, by Sectors, 1949-59

Industry	Number of Enterprises
Iron and steel	7
Nonferrous metals	14
Electric power stations*	24
Machinery	63
Coal	27
Oil	2
Chemicals	5
Drugs and pharmaceuticals	2
Paper	1
Textiles	1
Water conservation	1
State farms	1
Transportation	5
Others	3
Total	156

*Two of which manufactured petroleum machinery.

Source: Chu-yuan Cheng, Economic Relations Between Peking and Moscow (New York: Praeger Publishers, 1964), p. 28.

The precise value of the petroleum equipment furnished to China by the Soviet Union is not known, however, because some of the oil equipment delivered to China may not have been included in the category specified as petroleum equipment but instead in the "complete plant" category. Thus the statistics compiled by the U.S. government in early 1975 are not completely in agreement with the statistics given by K. C. Yeh in 1962.[6] According to the U.S. government, Chinese imports of equipment for geological survey, engineering, and gas extraction from the Soviet Union between 1955 and 1971 were as follows:[7]

Year	Imports (in thousands of dollars)	Year	Imports (in thousands of dollars)
1955	13,159	1963	113
1956	19,506	1964	754
1957	12,790	1965	863
1958	10,890	1966	1,697
1959	7,601	1967	530
1960	6,221	1968	330
1961	689	1969	171
1962	2	1970	166
		1971	99

A more detailed survey of the importing of oil equipment other than what was included in the "complete plant" category for the 1955-60 period has been provided by K. C. Yeh and is reproduced in Table 6.2.

Of the total machinery and equipment imported from the Soviet Union, petroleum equipment accounted for 8.2 percent in 1955, 8.7 percent in 1956, 5.4 percent in 1957, 6.9 percent in 1958, 3.7 percent in 1959, and 17 percent in 1960. In value terms, imports of oil-drilling equipment reached their peak in 1956 and have since declined steadily, indicating that domestic production has gradually replaced imported items.

China received foreign aid from Romania as well as from the Soviet Union. Romania provided China with petroleum equipment and technical data on the production of petroleum equipment. Together with the Soviet Union, these countries supplied China with 70,000 items of modern petroleum machinery and scientific instrumentation and over 2 million tons of supplies and materials during 1950-59.[8]

The sudden withdrawal of the Soviet technicians working in China in August 1960 marked the turning point of Sino-Soviet economic relations. Subsequently, imports of machinery and equipment from the Soviet Union dropped precipitously, and by 1962 China's imports of petroleum machinery and equipment from the USSR were reduced to only $2,000.

After Khrushchev's political demise in 1964, there was a brief hiatus in the Sino-Soviet schism. Between 1964 and 1966 China purchased $3.32 million worth of petroleum drilling equipment and $14.28 million of steel pipes from the Soviet Union, about one-fifth of China's annual imports during the 1950s.

PETROLEUM EQUIPMENT IMPORTED FROM WESTERN EUROPE AND JAPAN

To fill the vacuum created by the cessation of the Soviet supply of petroequipment, China turned to Western Europe and to Japan. In 1963, when Tach'ing was put into full operation, China ordered refining equipment and petrochemical plants from Western European countries.

The first set of oil-refining equipment from Western Europe was supplied by the Italian State Petroleum Company (Ente Nazionale Idrocarburi), or (ENI). A contract between Italy and China was signed in December 1963 for an integrated refining plant with an annual capacity of 300,000 tons, to cost $8.9 million. In the following year China signed a contract for two more plants, a refinery, and a plant to produce seamless steel pipe.[9]

TABLE 6.2

Imports of Oil Well Drilling Equipment and Pipes from the Soviet Union, 1955–60

Item	Imports (in millions of old rubles*)					
	1955	1956	1957	1958	1959	1960
Oil well drilling equipment	52.7	76.6	51.2	43.6	30.4	22.5
Oil well rigs	1.9	2.9	9.9	26.6	12.7	0.9
Drilling installations	27.2	30.9	11.8	7.5	12.7	9.5
Turbodrills	--	6.7	2.4	0.2	0.9	1.9
Casehardened parts	0.7	2.7	4.0	0.6	2.0	0.7
Chisel drill points	3.1	6.1	6.9	2.4		
Rockers	0.6	0.2	0.8	--		
Deep-well pumps	0.1	0.7	1.2	--	1.7	9.5
Seismic plants	2.3	4.5	--	1.7		
Core sampling plants	1.3	4.6	4.3	0.1		
Others	15.5	17.3	9.9	4.5	0.4	--
Pipes	19.6	29.0	7.2	45.0	57.5	54.4
Couplings for drilling pipes	2.9	1.6	0.2	--	--	--
Total	75.2	107.2	58.6	88.6	87.9	76.9

*One old ruble equals $.25 or .975 yuan.

Source: K. C. Yeh, Communist China's Petroleum Situation (Santa Monica, Calif.: Rand Corporation, 1962), p. 23.

Following Italy, West Germany in July 1964 sold China an integrated plant for heavy oil-cracking and olefin separation, at a price of $12.5 million. West Germany also supplied China with a plant to produce steel pipe, with an annual capacity of 40,000 tons, for $11 million.[10] In November 1972 the DEMAG compressor technology division of Duisburg shipped compressors for refineries and other petrochemical plants to China for 16 million deutsche marks. In January 1974, this West German firm negotiated with China one of the largest orders for compressors ever awarded in foreign trade for a sum of 31 million deutsche marks.[11]

In 1965 the French company Besliet was awarded a contract by the Chinese government to export oil-drilling equipment worth 24 million francs ($4.86 million) to China. In 1967 a British firm, Lowey Engineering Ltd., sold China steel pipe valued at $11 million.

Up to 1967, imports of petroleum equipment from Western Europe were rather meager; nor had China imported much from Japan up to that time. Prior to 1972 the government-financed Japan Export-Import Bank denied credits for trade with China. In the 1967-69 period, under the stress of the Cultural Revolution, the Chinese economy suffered severe disruptions. As capital investment declined, so too did imports of machinery and equipment. The total imports of all machinery and equipment, from the USSR and the non-communist countries combined, amounted to $267.8 million but 80 percent of their value in 1966.[12] In this period imports were reduced to a mere trickle.

The trend, however, was reversed in 1970, when the Chinese economy resumed its upward thrust and the Chinese petroleum industry commenced offshore exploration. Purchases of offshore drilling equipment dominated the shopping list as large contracts were awarded to companies in Japan, Holland, and Denmark.

As early as 1969, when the oil reserve underneath the Gulf of Pohai was confirmed, China ordered through the Toko Bussan, of the Ishikawajima-Harima Heavy Industries of Japan, a sea-bed drilling rig valued at 4,000 million yen. In addition, China also submitted a request for a bid from Mitsubishi Heavy Industries for the construction of another offshore drilling rig, which incorporated U.S. technology. The order as well as the negotiations were canceled in April 1970 as a consequence of the "four conditions" laid down by Chou En-lai regarding Japanese firms that traded with Taiwan. In June 1970 China placed a new order to Nippon Kaiyo Kussaku (Japan Offshore Drilling Co.), a subsidiary of Mitsubishi, and in September 1972 Japan exported to China the Fuji, an offshore well-drilling rig, with its support ship and accoutrements, at a cost of 2,600 million yen ($9.8 million). The Fuji is capable of drilling 15,000 feet in a maximum of 175 feet of water and is now operating in the Pohai Bay.[13]

In early 1973, as China escalated her efforts to exploit the offshore oil resources, the president of the Asia Offshore Drilling
Company of Japan was invited to Peking to negotiate the sale of a
"jacket-type" undersea oil-drilling unit, a stationary rig suitable for
use in shallow waters of the Pohai Gulf. The negotiation was followed by a discussion between Chinese government officials and the
Nippon Kokan Company, Japan's leading pipeline company, for the
sale of onshore oil pipelines 60 centimeters in diameter, to be a part
of the 715-mile pipeline connecting Tach'ing and Chinwangtao.[14] In
June 1973 Nippon Kokan was awarded a contract to build eight self-
propelling bucket dredgers to be delivered in 1974-75, for a total
cost of 14,000 million yen ($53 million). The dredgers are to be
used in conjunction with the Pohai Gulf development. By invitation
of the China National Technical Import Corporation, eight Japanese
technical experts, representing Nippon Steel Company, Mitsui Company, Mitsui Ocean Development and Engineering Corporation, and
Teikoku Oil Company arrived in Canton in October 1973 for a discussion of supplying China 120-centimeter (48-inch) diameter offshore
pipeline to be laid in the Pohai Gulf. Two months later China awarded
a contract of 6,000 million yen ($21.4 million) to Mitsubishi Heavy
Industries for the Number 2 Hakuryu, a secondhand heavy-duty drilling rig, with option to purchase a second rig.[15]

In addition to the procurement activities with Japan, China also
acquired offshore drilling facilities from the Northern European
countries. In early 1973 a $39.3 million contract was granted to the
N.V. Industrieele Handelscombinate of The Netherlands for trailing
suction hopper dredges to be used in Pohai. In December 1973 China
ordered eight 160-foot offshore supply vessels from the Weco shipping company of Denmark for a price of $20 million. The vessels,
which can service up to 10 rigs, are scheduled to be delivered between January 1974 and October 1975.

One 1974 report from Hong Kong reveals that China has re-
quested the Nippon Steel Company, the Mitsui Ocean Development
and Engineering Corporation of Japan, and the Italian State Petroleum Company (ENI) to discuss the construction of submarine pipelines that would link a recently confirmed offshore producing well to
shore facilities.[16]

Chinese overseas petroequipment acquisitions since 1972 provide cogent evidence that China considers the exploitation of oil resources under the continental shelf to be the most promising path for
the development of the Chinese petroleum industry. The decision to
obtain equipment and technology from Japan and from the West potentially opens an avenue for U.S. trade, since U.S. petroequipment and
technology is superior to that of Western Europe and Japan.

TABLE 6.3

Imports of Oil Well Drilling and Refining Equipment and Pipes from Europe and the United States, 1963-73

Firm	Equipment	Value (in millions of U.S. dollars)	Date Contract Was Signed
Italy			
ENI	oil refinery plant[a]	8.9	12/63
Snam Projetti	oil refinery plant	5.6	6/65
ENI	oil refinery plant	NA	9/65
Innocenti	seamless steel pipe	3.2	9/65
France			
Berliet	oil drilling equipment	4.86 million francs	1965
West Germany			
Lurgi-Gesellschaft	heavy oil cracking and olefin separators[b]	12.5	7/64
Mannesmann AG and Lowey Eng.	steel pipe	11.0	1967
DEMAG	compressors	16 million marks	11/72
DEMAG	compressors	31 million marks	1/74
Netherlands			
N. V. Industreele Handelscombinate	four pieces trailing equipment suction hopper dredges	39.3	1/73
Denmark			
Weco Shipping Co.	eight oil rig supply and towing vessels	20.0	1/73
United States			
Rucker	two land blowout preventer stacks	2.0	1/73

NA = Not available.

[a]300,000 tons per year.

[b]50,000 tons per year, located at Lanchow.

Sources: (1) Ajia Keizai Jumpo [Asia Economic Thrice-monthly] (Tokyo), late June 1964, pp. 11-14; (2) U.S. China Business Review, January-February 1974, p. 31; (3) New York Times, March 12, 1974; (4) China Trade Report (Hong Kong), various issues; (5) Yearbook on Chinese Communism 1973 (Taipei: Institute for the Study of Chinese Communist Problems, 1973), Section 5, pp. 56-58.

TABLE 6.4

Petroleum Equipment Contracted from Japan, 1972–75

Firms	Type of Equipment	Value (in millions of U.S. dollars)	Date Contract Was Signed
Japan Offshore Drilling Company	offshore well-drilling rig "Fuji" (capacity 4,650 meters in a maximum of 53 meters of water)	9.8	September 1972
Hakodate Dockyard	6 dredgers and 8 sludge carriers		November 1972
Nippon Kokan	8 self-propelling bucket dredges	53	June 1973
Mitsubishi Heavy Industries	number 2 Hakuryu second-hand heavy-duty drilling rig*	21.4	December 1973
Asia Offshore Drilling Company	jacket-type undersea oil drilling unit		under negotiation
Nippon Kokan	onshore oil pipeline (60 centimeters in diameter)		under negotiation
Nippon Steel, Mitsui & Company, Mitsui Ocean Development of Engineering Corporation, and Teikoku Oil Company	offshore pipeline (120 centimeters in diameter)		under negotiation
Mitsubishi Heavy Industries	offshore drilling rig number 2 Hakuryu (maximum 20 meters of water)	22.6	December 1973
Hitachi Shipbuilding and Engineering	supply boats (diesel) (5)--capacity (each) 600 tons	10.0	September 1973
Sumitomo Shoji Ocean Systems	tugboats (2)--capacity (each) 9,000 h.p., underseas survey craft (2), and 500-ton survey ship		under negotiation

*For Pohai Gulf.

Sources: (1) Pacific Community 5, no. 3 (1974): 466-67; (2) U.S.-China Business Review, January-February 1974, p. 31; (3) Far Eastern Economic Review, various issues; (4) Hans Heymann, Jr., "Acquisitions and Diffusion of Technology in China," in Joint Economic Committee, China: A Reassessment of the Economy (Washington, D.C.: Government Printing Office, 1975), pp. 718-19.

PROCUREMENT OF PETROCHEMICAL PLANTS

Equally important in China's procurement list are the petrochemical plants. Since 1962 China's developmental policy has placed high priority on agricultural progress. The expansion of chemical fertilizer production constitutes an essential ingredient in the new policy. With the rapid advance of petroleum industry, the country now possesses adequate raw materials for manufacturing large quantities of chemical fertilizer; the primary constraint is the supply of machinery and equipment. For this reason the purchase of petrochemical plants has become a major import item for China during the past decade.

In October 1963 China purchased for £3 million ($8.4 million) its first integrated synthetic ammonia plant from the United Kingdom. The plant was built at Luchow in Szechwan and is capable of producing 160,000 tons per year. This was followed by a contract in September 1964 to import an integrated plant for the manufacturing of high pressure polyethylene from the United Kingdom for £4.5 million ($12.6 million). Over the following three years China ordered another four petrochemical plants from the United Kingdom at a total cost of £14 million ($39.2 million).

Italy, apart from supplying China with refineries, also supplied two urea plants with a total capacity of 600,000 tons per year, costing China $14.2 million. The contracts were signed in December 1963 at Peking.

In 1963 France furnished China with two plants for making ethylene, hexanol, and butanol at a price of 42 million francs ($7.2 million), while West Germany exported two chemical plants to China in 1964-65 at a total cost of $12.75 million, one of which produces perlon (synthetic fiber), and the other, acrylenitrite.

The Netherlands and Norway, among other European countries, also provided China with petrochemical plants in the 1963-67 period. The former sold China three urea synthetic plants for $10.5 million; the latter, a naphtha cracking plant for $14 million.

Large-scale acquisition of Western and Japanese equipment occurred primarily after 1970, however, with a disproportionate share of the orders placed with Japanese firms. As early as 1963 China had contracted the Kurashiki Rayon Company of Japan to supply an integrated vinylon plant. The contract for the $20 million plant was signed in June 1963, but was not executed because the Japanese government refused to grant the required export credits to the Kurashiki Rayon Company.

The Japanese government's policy toward trade with China changed dramatically in 1972, and subsequently it has offered China long-term credit at low interest rates. As a consequence, exports of petrochemical plants from Japan to China experienced a tremendous increase. In the 1972-73 biennium, 14 contracts were awarded to Japanese firms with an aggregate value of over 400 million dollars.

The largest and most impressive single contract, however, was granted to a French consortium in September 1973. Led by Technip and Speichem, this consortium garnered $300 million (1.2 billion French francs) to supply the equipment for a large petrochemical complex being built 60 kilometers south of Shenyang (Mukden) in Southern Manchuria. The entire project involves 17 plants (see Figure 6.1).

The entire complex, when completed, will produce 2 million tons of petrochemical and synthetic fibers. It will rank as China's most important producer of petrochemicals.[17]

From the perspective of U.S. trade, a most significant development was the recent order by the Chinese government to M. W. Kellogg, a division of Pullman, Inc., for eight ammonia plants. This $205 million contract was the largest order from China for petrochemical machinery landed by a U.S. firm.[18]

TABLE 6.5

Contracts of Petroleum Exploration, Extraction, and
Petrochemical Plants from Abroad, 1973-75
(costs in millions of U.S. dollars)

	Number of Units	Estimated Cost
Petroleum exploration and extraction plants		127
Offshore drilling platforms	4	
Oil rigs	2	
Survey and supply vessels	33	
Petrochemical and synthetic fiber plants		900
Intermediate product	33	
Synthetic fibers	11	
Total	83	1,027

Note: This table excludes chemical fertilizer plants which amounted to £534 million during the same period.

Sources: Hans Heymann, Jr., "Acquisition and Diffusion of Technology in China," in Joint Economic Committee, China: A Reassessment of the Economy (Washington, D.C.: Government Printing Office, 1975), p. 701.

FIGURE 6.1

Flow Chart for the Petrochemical and Synthetic Fiber Complex in Southern Manchuria

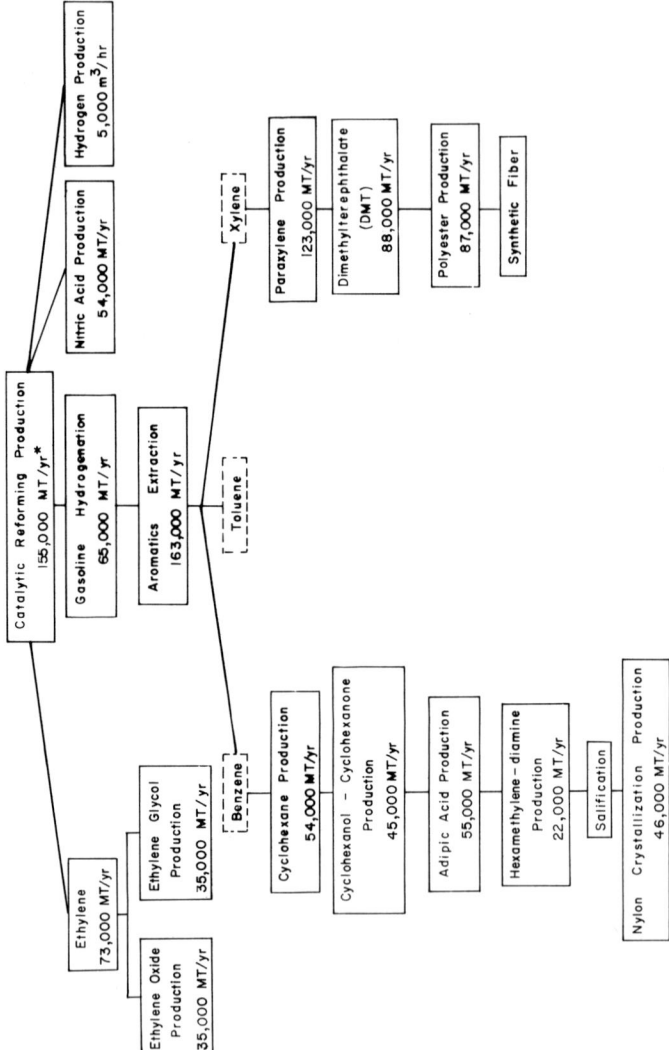

*MT/yr = metric tons per year.

Source: Hans Heymann, Jr., "Acquisitions and Diffusion of Technology in China," in Joint Economic Committee, China: A Reassessment of the Economy (Washington, D.C.: Government Printing Office, 1975), pp. 718-19.

TABLE 6.6

Imports of Petrochemical Plants from the West, 1963-75

Firm	Type of Equipment	Value (in millions of U.S. dollars)	Date Contract Was Signed
United Kingdom			
Humphrey and Glasgow	ammonia[a]	8.4	October 1963
Simon Carves	high pressure polyethylene	12.6	September 1964
Vickers Zwimmer	polypropylene	7	March 1965 (canceled July 1968)
Humphrey and Glasgow	ammonia[b]	23.8	August 1965
Prine	acrylic resin[c]	8.4	September 1965
West Germany			
Friedrich Uhde Gmb H.	perlon (synthetic fiber)[d]	1.75	July 1964
Lurgi-Gesellschaft	acrylenitrite[e]	11	May 1965
Friedrich Uhde and Hoechst	acetaldehyde[f]	4	July 1973
Uhde	vinyl chloride monomer	19	January 1974
Uhde	polyethylene	15	March 1974
Italy			
Monticattini	urea[g]	14.2	December 1963
Snam Projetti	aromatics chemical[h]	5.55	1966
Snam Projetti	polypropylene[i]	16	January 1974
France			
Melle and Speichem	ethylene, hexanol, and butanol[g]	FF. 8.5 million	January 1964
Speichem	vinyl acetate and methanal	90	May 1973
Technip and Speichem	petrochemical complex	282	September 1973
Heurtey Industries	ammonia and urea complexes[j]	120	February 1974
Rhone Poulenc Textiles	nylon spinning	10	August 1974
Netherlands			
Oriental Engineering	urea[k]	2.5	September 1963
Continental Engineering	urea[l]	6 to 8	1965
Kellogg Continental[m]	urea[j]	34	February 1973
Kellogg Continental[m]	urea[n]	55	September 1973
Norway			
	naphtha cracking	14	July 1965
United States			
M. W. Kellogg	ammonia[o]	73	March 1973
M. W. Kellogg	ammonia[n]	130	November 1973

[a]160,000 tons per year, located at LuChow.
[b]Four plants.
[c]Located at Lanchow.
[d]Located at Shanghai.
[e]10,000 tons per year.
[f]30,000 tons per year.
[g]Two plants with a total of 300,000 tons per year.
[h]70,000 tons per year.
[i]35,000 tons per year.
[j]Three plants.
[k]100,000 tons per year.
[l]Two plants with a total of 175,000 tons per year.
[m]Subsidiary of M. W. Kellogg, Houston, Texas.
[n]Five plants.
[o]Three plants, probably feedstock for the Dutch urea plants.

Sources: (1) Central Intelligence Agency, People's Republic of China: International Trade Handbook, September 1974, pp. 16-17; (2) Yearbook on Chinese Communism 1973 (Taipei: Institute for the Study of Chinese Communist Problems, 1973), Section 5, pp. 56-58; (3) Ajia Keizai Jumpo [Asia Economic Thrice-monthly] (Tokyo), late June 1964, pp. 11-14; (4) Japan External Trade Promotion Organization, China News Letter, March 1974; (5) U.S. China Business Review, January-February 1974, pp. 36-37.

TABLE 6.7

Imports of Petrochemical Plants from Japan, 1964-74

Firm	Type of Equipment	Value (in millions of U.S. dollars)	Date Contract Was Signed
NA	acetylene generation[a]	3	May 1964
Toyo Engineering and Mitsui Toatsu Chemical	ethylene[b] and butadiene[c]	46	December 1972
Mitsubishi Yuka and Mitsubishi Heavy Industries	ethylene[d] and poval	34	February 1973
Asahi Kasei	acrylonitrile monomer[e]	30	March 1973
Kuraray Industries	vinyl acetate[f] and poval[g]	26	March 1973
Mitsui Toatsu Chemical	ammonia[h] and urea[a]	42	April 1973
Toray and Mitsui Shipbuilding	polyester polymerization[j]	50	May 1973
Sumitomo Chemical	benzene, toluene, and xylene[k]	5	May 1973
Mitsubishi Petrochemical	polyethylene, high pressure, low density[l]	22	July 1973
Sumitomo Chemical	polyethylene, high pressure, low density[m]	41	August 1973
Mitsui Toatsu Chemical and Toyo Engineering	ammonia[h] and urea[n]	43	September 1973
Mitsui Petrochemical and Mitsui Shipbuilding and Engineering	polypropylene[o]	25	October 1973
Nisso Petrochemical	ethylene glycol	15	December 1973
Toho Titanium	polypropylene catalyst[p]	5	January 1974
Kuraray	polyvinyl alcohol	19	February 1974

NA = Not available.

[a] 1,100 cubic meters per year.
[b] 300,000 tons per year.
[c] 45,000 tons per year.
[d] 120,000 tons per year.
[e] 50,000 tons per year of acrylonitrile; 1,000 tons per year of acetonitrile; 5,000 tons per year of cyanic acid.
[f] 66,000 tons per year.
[g] 33,000 tons per year.
[h] 330,000 tons per year.
[i] 528,000 tons per year.
[j] 25,000 tons per year.
[k] 50,000 to 60,000 tons per year.
[l] 60,000 tons per year.
[m] 180,000 tons per year.
[n] 30,000 tons per year.
[o] 80,000 tons per year.
[p] For polypropylene plant bought from Mitsui.

Sources: (1) Japan External Trade Promotion Organization, China News Letter, March 1974; (2) Central Intelligence Agency, People's Republic of China International Trade Handbook, September 1974, pp. 16-17; (3) Japan Times, various issues.

Total Chinese contracts of complete plants in the fields of petroleum exploration and extraction, petrochemical and synthetic fiber during the period from January 1973 to March 1975 amounted to approximately $1 billion, the distributions of which are given in Table 6.5. The procurement of petrochemical plants continued into 1974, but at a reduced pace. In the first eight months of 1974, contracts were let for petrochemical plants valued at $194 million, as against the $1,077 million contracted during the calendar year of 1973. (See Tables 6.6 and 6.7.)

THE ACQUISITION OF FOREIGN TECHNOLOGY

This extensive procurement of foreign equipment for the petroleum and petrochemical industries, has been paralleled by concerted efforts by Chinese officials to accelerate the acquisition of the latest technology from the West and Japan. Since the efficient use of sophisticated petroleum and petrochemical equipment requires a high level of technology, it is not surprising that there has been an increased flow of technical experts between China and the more industrialized countries.

Since 1972 Chinese technical delegations have been dispatched to tour the principal oil-producing nations in the West in order to study their experiences in petroleum drilling, refining, and petrochemical production. In May 1972 the Chinese Petroleum Exploration and Exploitation Corporation sent a delegation to Iran to negotiate imports of Western technology. In September 1972 a 17-man Chinese petroleum study group led by T'ang K'e, Vice-Minister of the Fuel and Chemical Industries, visited Canada. The group spent three weeks in Canada studying oil refineries and gas plants and evaluating equipment manufacturers. They also visited offshore drilling rigs being constructed in Nova Scotia, as well as some in operation.[19] The same study group subsequently traveled to France for a four-week inspection of French petroleum equipment and refining facilities. In March 1973 a seven-member delegation from the China Oil and Natural Gas Development and Exploration Corporation arrived in London to take a closer look at the installations of Shell and the British Petroleum Company. In May of 1973 the prominent authority on high polymer, Ch'ien Jen-yuan, and a delegation of six technical experts visited Japan.[20]

Foreign experts were also invited to China. In June 1971 an Iraqi economic and technical delegation led by Su'dum Hammadi, the Minister of Oil and Mining, arrived in Peking. Among the members of the delegation were the head of research of the Iraqi National Minerals Company and the deputy director of the managing board of the Iraq National Oil Company. An agreement on economic and technical cooperation between China and Iraq was signed in Peking on June 21, 1971. Since the Iraqi oil industry is well developed, the presence of a top Iraqi mineral researcher and a leading Iraqi oil prospective geologist at the talks and the signing of an agreement suggests the possibility that Iraq may provide help in exploiting China's oil resources.[21]

A second method utilized by China to introduce foreign technology was to induce advanced industrial countries to participate in industrial exhibitions in China. Between 1972 and 1973 almost all the major trade partners in Western Europe held industrial exhibitions in Peking. By the end of 1974, no less than 32 have been held, with six more planned for 1975.[22] At the same time, the Chinese government mobilized technicians and college students in order to obtain detailed drawings of the machinery on display as well as to engage in scientific exchanges with the representatives of the major corporations present. Even the U.S. firms invited to the Spring Export Commodities Fair in Canton were subjected to considerable interrogation on a variety of American technological processes.[23]

However, the most direct way to borrow foreign technology is to make it part of an equipment contract. With respect to integrated-plant purchase agreements, arrangements for the training of technicians were frequently provided in the contract. In some cases arrangements were made for a five-year exchange of know-how, as well as for the training of Chinese engineers in the equipment-producing countries. In some contracts, as in the case of the purchasing of acrylonitrile and associated plants from the Asahi Chemical Industrial Company in Japan, 20 percent of the contract value was allocated for technical services.[24] The contracts with M. W. Kellogg Company of Houston for supplying China with eight ammonia plants, signed in 1973, specified the design, equipment, and engineering services required. Chinese technicians have since visited the Houston facilities in connection with these contracts.[25]

THE EMERGING PATTERNS

China's petroleum equipment procurement since 1949 has some discernible patterns.

First, as exhibited in Tables 6.2 through 6.7 in the preceding sections, the composition of the equipment China imported from abroad changed from time to time. In the 1950s most equipment was bought from the Soviet Union and consisted of onshore oil well drilling equipment, which accounted for 70 percent of the total imports in 1955, 71 percent in 1956, and 88 percent in 1957. (See Table 6.2.)

Subsequently, imports of drilling machines declined steadily, being 49 percent of the total petroleum equipment imported in 1958, 35 percent in 1959, and 29 percent in 1960. This would indicate that by 1960 China could provide most of its own drilling machines for on-shore exploration.

In the 1960s, as shown in Table 6.3, the most important items imported from overseas were oil refinery plants. Since 1970, when China started exploiting offshore oil resources, offshore drilling rigs and supporting vessels have dominated the procurement list. However, as China's crude oil output multiplied, the country also began to expand her petrochemical industries. Imports of petrochemical plants, which commenced in the mid-1960s, peaked in 1973, when more than 1 billion dollars worth of petrochemical plants were contracted. From the changing pattern of equipment procurement, it would appear that the long-term objective of the Chinese petroleum industry is not to export crude oil in huge quantities, as the OPEC countries have tended to do in the past, but rather to expand China's petrochemical industry. An analysis of the major factors involved suggests that during the 1980s petrochemical industry may become one of the most significant industries in China.

The second discernible pattern in China's petroleum equipment procurement has been in the partners from whom China imported her equipment, which have also shifted with time. Until the Sino-Soviet split in 1960, the Soviet Union was the chief supplier and provided 96 percent of the petroleum equipment that China imported between 1950-59. In the 1960s Western Europe replaced the Soviet Union as the principal supplier of oil equipment, but since 1970 Japan has led the foreign suppliers, not only with respect to offshore drilling equipment, but also by obtaining the largest share of contracts let for petrochemical plants. Of the $1,077 million contracted for foreign petrochemical plants in 1973, $389 million, or 36 percent, went to Japan. In recent years, however, the United States has again been initiated into the ranks of the major trade partners of China; in 1973 the United States exported $205 million in petrochemical plants to China, 20 percent of the year's total of Chinese petroequipment imports.

The change in trade partners is partly the result of changing domestic demands, but it is also dictated by changing international relations. When the People's Republic of China pledged the "lean to one side" foreign policy in the early years, the Soviet Union was the sole source of supply. In the 1960s, when China was at odds with both the USSR and the United States, the European Common Market and Japan became the natural alternatives. The movement toward normalization of the relationships between China and the United States in recent years has helped to promote U.S. exports to China.[26]

Finally, China's foreign trade policies have changed in recent years. In the past China adhered to the principle of balanced trade and declined to accept credits offered by noncommunist countries. The ability to import was therefore directly constrained by her ability to export in a given year. Since 1949, except during the 1951-55 period, when China accepted Soviet credit to finance her trade deficit, China has managed to maintain a trade surplus. Since 1972 a new pattern appears to be emerging with regard to the use of the deferred-payments mechanism with Japan. In August 1972, in a discussion with a Japanese representative, Chinese Premier Chou En-lai stated that although China would not accept charity exports backed by government assistance from foreign countries, China would welcome exports based on international payments standards.[27] This was the first time that a Chinese official had indicated acceptance of the concept of deferred payments for imports. As of 1973, most export credits granted by Japan were at an annual interest rate of 6 percent, with a repayment period of about five years. The acceptance of deferred payment has considerably enlarged China's ability to import. The large-scale procurement of integrated plants in 1973 may reflect this new policy orientation and provide a pattern for future trade relations.

NOTES

1. Mao Tse-tung, Selected Works, vol. 4 (Peking: Foreign Languages Press, 1961), p. 417.

2. Chu-yuan Cheng, Economic Relations between Peking and Moscow (New York: Praeger Publishers, 1964), pp. 1-5.

3. Wang Chih-ch'un, "The Petroleum Industry in the People's Era," Chung-Kuo Hsin-wen, September 12, 1959.

4. Cheng, op. cit., p. 43.

5. Chu-yuan Cheng, The Machine-building Industry in Communist China (Chicago: Aldine-Atherton, 1971), p. 172.

6. K. C. Yeh, Communist China's Petroleum Situation (Santa Monica, Calif.: Rand Corporation, 1962).

7. Central Intelligence Agency, People's Republic of China Foreign Trade in Machinery and Transportation Equipment since 1952 (Washington, D.C.: Government Printing Office, 1975), p. 13.

8. Wang Chih-ch'un, op. cit., p. 53.

9. China Trade News, London, April 1968.

10. Far Eastern Economic Review, September 23, 1965, p. 566.

11. Euro-Commerz, West Berlin, January 1974, p. 22.

12. Central Intelligence Agency, People's Republic of China International Trade Handbook, December 1972, p. 25.

13. Yoshio Koide, "China's Crude Oil Production," Pacific Community 5, no. 3 (April 1974): 467.

14. Far Eastern Economic Review, July 23, 1973, p. 32.

15. U.S. China Business Review, January-February 1974, p. 31.

16. New York Times, March 12, 1974, pp. 49 and 54.

17. Euro-Commerz (West Berlin), January 1974, p. 22; also U.S.-China Business Review, January-February 1974, p. 37.

18. New York Times, November 28, 1973.

19. China Trade Report (Hong Kong), September 1972, p. 2.

20. Yearbook on Chinese Communism 1973 (Taipei: Institute for the Study of Chinese Communist Problems, 1974), section 5, pp. 48-49.

21. China Trade Report (Hong Kong), June 1971, p. 3.

22. Hans Heymann, Jr., "Acquisition and Diffusion of Technology in China," in Joint Economic Committee, China: A Reassessment of the Economy (Washington, D.C.: Government Printing Office, 1975), p. 699.

23. China Trade Report (Hong Kong), September 1972, pp. 3-4.

24. Japan External Trade Promotion Organization, China Newsletter, March 1974, pp. 15-16.

25. U.S. China Business Review, Washington, D.C., March-April 1974, p. 56.

26. Alexander Eckstein, "China's Economic Growth and Foreign Trade," U.S. China Business Review, July-August 1974, pp. 15-20.

27. China Trade Report (Hong Kong), August 1972, pp. 3-4.

CONTRIBUTIONS TO THE
NATIONAL ECONOMY

The importance of the petroleum industry to the Chinese national economy was relatively small during the early stages of industrialization but it has become increasingly strategic in recent decades. The varying impact of oil production on the economies of the oil-producing countries themselves can be analyzed by studying the historical records of these countries, be they in North America, Latin America, the Middle East, or Eurasia; by and large the net impact appears to be a function of the relative share of oil production to the gross national product, industrial employment, and the balance of payments. In countries where oil is used extensively both as a raw material and for fuel for domestic development, the most profound impact may arise from its linkage effects.

Since the mid-1960s the concept of backward and forward linkages has been used to explore the interdependence of various industries. According to Albert O. Hirschman, two kinds of interdependence or linkage effects on industry need to be explored: the input provision, derived demand, or backward linkage effect; and the output utilization, or forward linkage effect.[1] Sometimes these linkage effects have been euphemistically wrapped in the accolade "power of dispersion." Thus, if expansion of a certain industry leads to a general increase in economic activity embracing all or at least most industries, such an industry must be classified as a "key industry" and merits a high priority in development.[2]

By utilizing the concept of linkage effects, it is possible to assess the contribution of the petroleum industry in a more meaningful perspective. Oil and natural gas from oil fields not only provide an efficient fuel for industry, transportation, and agriculture, but also supply the most important basic raw material for the petrochemical industry. Even within the iron and steel industries, the development of a process whereby fuel oil or natural gas can be injected into blast furnaces may partially eliminate the dependence on coke and thereby foster industrial growth in iron and steel in countries lacking coal. Moreover, the expansion of the oil industry also stimulates the demand for steel products, especially steel tubes and pipes. Furthermore, since one-third to one-half of the investment of the petroleum industry is for the acquisition of machinery and equipment, the expansion of the oil industry provides a ready market for the petroleum equipment industry.[3]

As the Chinese policy with respect to the petroleum industry is oriented toward both the expansion of exports of crude oil and the development of the petrochemical industry, this chapter will examine the direct and the backward and forward linkage effects.

THE CONTRIBUTION TO GROSS INDUSTRIAL
OUTPUT VALUE

The direct contribution of the petroleum industry can be assessed through its relative contribution to the gross national product. Since we lack national income accounting data for the Chinese economy during the 1960s and 1970s, the alternative is to measure the relative value of petroleum output versus the value of the gross industrial output, for which some rough estimates can be made.

The only official figure released denoting the gross output value of China's petroleum industry was for the year 1956, when the petroleum industry reportedly produced a gross output value of 650 million yuan.[4] Using this figure as a base, and assuming that the gross output value of the petroleum industry has risen in proportion to the annual growth of crude oil output, a series of gross output value figures for various years can be derived for the years 1952-74, as shown in Table 7.1. The validity of this derivation depends on the assumption that the value of crude oil and petroleum output grew proportionately over the entire time span, however. Unfortunately, this does not hold in the long run.

According to official sources, crude oil output rose 8.54 times between 1949 and 1956 and the gross output value of the petroleum industry increased 21 times.[5] These figures imply that the gross output value of the petroleum industry had risen at a rate about 2.4 times as great as that of crude oil output. This divergency can be attributed to the fact that China's crude oil output lagged behind refining capacity during the First Five-Year Plan period (1953-57), which is evidenced by China's importation of crude oil during this time. However, this trend was reversed in the 1960s and 1970s. Following the opening of the eastern oil fields, crude oil output expanded at a rate much faster than refining capacity. Between 1965 and 1974 crude oil output

TABLE 7.1

Estimated Gross Output Value in the
Petroleum Industry, 1952-74

Year	Crude Oil Output (in thousands of tons)	Index (1956 = 100)	Gross Output Value of Petroleum Industry (in millions of yuan at 1952 prices)
1952	436	37	240
1953	622	53	340
1954	789	67	440
1955	966	93	600
1956	1,163	100	650
1957	1,458	125	810
1958	2,264	194	1,260
1959	3,700	318	2,070
1960	5,200	447	2,910
1961	6,000	515	3,350
1962	6,700	576	3,740
1963	7,500	644	4,190
1964	8,500	730	4,750
1965	11,000	945	6,140
1966	14,000	1,203	7,820
1967	11,000	945	6,140
1968	15,000	1,324	8,610
1969	20,700	1,779	11,560
1970	29,100	2,502	16,260
1971	37,500	3,224	20,960
1972	45,000	3,859	25,080
1973	53,000	4,557	29,620
1974	64,000	5,503	35,770

Sources: Crude oil output as computed by the author is shown in Tables 2.1 through 2.5 (sources are given in Tables 2.1, 2.2, and 2.4); gross output value is computed using the 650 million yuan in 1956, from Chao I-wen, Hsin-chung-kuo ti Kung-yeh (Peking: T'ung-chi Ch'u-pan-she, 1957), p. 50, as basis and applying it to the crude oil output index in various years.

increased fourfold, whereas refining products increased twice. [6] The use of crude oil output indices to derive the gross value of petroleum output may thus involve significant overestimation. However, this upward bias may be offset by other factors. As China's petroleum industry matures, the improvement in the quality of refining as well as the greater number of varieties of high-grade products tends to increase the per-ton value of output of petroleum products. While the net effect of these two opposing trends is difficult to determine, the bias imparted by the derivation is probably not great.

The gross output value of all industry for the period 1949-57 in 1952 constant prices was published in Ten Great Years. [7] Subsequently, no official statistics have been made available. However, from fragmentary information, the value of industrial output for 1963-66 can also be estimated. In early 1975, in his Report of Government Works to the National People's Congress, Chou En-lai announced that industrial output in 1974 amounted to 290 percent of that of 1964. [8] Given this information, the value of the gross output for several benchmark years can be derived and compared to the value of petroleum output as listed in Table 7.2.

The changing importance of the petroleum industry can be vividly illustrated by comparing its relative share of total output with that of seven other basic industries. As shown in Table 7.3, the petroleum industry share advanced from 1 percent in 1957 to 3.6 percent in 1965, 7.4 percent in 1972, and 8.2 percent in 1974. Between 1957 and 1974 the relative share of the power industry increased slightly, from 1.4 percent to 1.7 percent, while the relative share of coal declined from 2.2 percent to 1.1 percent. It is obvious that by 1974 the value of the petroleum industry output exceeded the value of the combined output of these two energy industries.

In only 17 years (1957-74), the petroleum industry has moved from the bottom of the list of these eight basic industries to a position third from the top, being surpassed only by the machine-building and chemical industries.

THE CONTRIBUTION TO EMPLOYMENT

The employment-generating effects of the petroleum industry are less impressive than its gross value of output. In the initial years (1950-52), petroleum employment was the lowest of the 12 major industrial sectors in China. Employment in the petroleum industry totaled 12,000 in 1949 and 22,000 in 1952, only .4 percent of China's total industrial employment for 1952. The increased investment in the petroleum industry during the First Five-Year Plan period (1953-57) was accompanied by a tripling of its employment.

TABLE 7.2

Petroleum Output Value as a Percentage of Total Gross Industrial Output Value

Year	Gross Output Value (in millions of yuan)		Petroleum as a Percentage of Total Industrial Output
	Total Industry	Petroleum Industry	
1952	34,330	240	0.60
1953	44,700	340	0.76
1954	51,970	440	0.80
1955	54,870	600	1.00
1956	70,360	650	0.90
1957	78,390	810	1.00
1963	131,200	4,190	3.10
1964	150,880	4,750	3.10
1965	167,480	6,140	3.60
1966	200,980	7,820	3.80
1972	334,960	25,080	7.40
1974	437,550	35,770	8.20

Sources: (1) Total industry figures for 1952-57 are from People's Republic of China, State Statistical Bureau, Ten Great Years (Peking: Foreign Language Press, 1960), p. 87. This series includes the output value of handicrafts and is calculated at 1952 constant prices.

(2) The 1963 industrial output value is derived through the following two steps: (a) According to Ta Kung Pao (Hong Kong), October 9, 1964, the gross value of handicrafts output was more than four times that of 1949. The 1949 handicrafts output value was given as 3,240 million yuan at 1952 prices (Ten Great Years, op. cit., p. 94). The 1963 output value of handicrafts can be calculated as 13,120 million yuan. (b) According to Jen-min Jih-pao, October 27, 1963, editorial, the gross output value of handicrafts in 1963 accounted for 10 percent of the gross output value of industry. Therefore, the 1963 industrial output value must be 131,200 million yuan.

(3) Total industry figures for 1964-65 are based on Chou En-lai's report to the first session of the Third National People's Congress on December 21, 1964, that industrial output rose 15 percent in 1964 and was expected to increase 11 percent in 1965; (4) the total industry figure for 1966 is from Jen-min Jih-pao, December 31, 1966, p. 2; (5) the total industry figure for 1972 assumes an increase of 100 percent between 1965 and 1962, as reported in Chu-yuan Cheng, "China's Industry: Advances and Dilemma" in Current History, September 1971, pp. 154-58; (6) the total industry figure for 1974 is from Peking Review, no. 4 (January 24, 1975), p. 22.

(7) The gross output value of the petroleum industry is computed in Table 7.1. (Sources are given in Tables 2.1, 2.2, 2.4, and 7.1.)

TABLE 7.3

Relative Shares of Eight Major Basic Industries
in Total Industrial Output Value, 1957-74

	1957	1964	1965	1972	1974
Total industrial output (in millions of yuan at 1952 prices)	78,390	150,880	167,480	334,960	437,550
Petroleum	810	4,750	6,140	25,080	35,770
Machinery	6,177	22,943	30,973	86,725	NA
Ferrous metals	5,202	9,723	10,696	22,364	21,391
Building materials	1,626	2,581	3,501	5,807	NA
Chemicals	4,291	9,103	13,731	30,548	NA
Power	1,105	2,554	2,977	6,716	7,662
Coal	1,779	2,716	2,988	4,848	5,188
Timber	1,151	1,403	1,485	1,914	NA
Total industrial output (in percentage)	100.0	100.0	100.0	100.0	100.0
Petroleum	1.0	3.1	3.6	7.4	8.2
Machinery	7.8	15.2	18.4	25.8	NA
Ferrous metals	6.6	6.4	6.3	6.6	4.8
Building materials	2.0	1.7	2.0	1.7	NA
Chemicals	5.4	6.0	8.1	9.1	NA
Power	1.4	1.6	1.7	2.0	1.7
Coal	2.2	1.9	1.7	1.4	1.1
Timber	1.4	0.9	0.8	0.5	NA

NA = Not available.

Sources: (1) Total industrial output value and petroleum output are given in Table 7.2; (2) other items for 1957-72 are estimated by Thomas G. Rawski in his paper, "Measuring China's Industrial Performance 1949-72" (presented at the Conference on Quantitative Measures of China's Economic Output, Washington, D.C., January 17-18, 1975), Table 15; (3) the 1974 output value for the ferrous metals, power, and coal industries are derived by the author based on Chou En-lai, "Report on the Work of the Government," Peking Review, January 24, 1975, p. 22: according to Chou, steel output in 1974 was 220 percent of that in 1964, while output of power in 1974 was 300 percent of that of 1964 and for coal 191 percent of that of 1964.

By the end of 1957 employment had reached 67,000. In 1958, under the impetus of the Great Leap Forward program, employment surged 85 percent, to an estimated 124,000.[9] However, the huge growth of employment in that year can partially be attributed to the thousands of small, ill-fated shale-oil refineries throughout the country.

Official statistics for petroleum employment are unavailable after 1958. Fragmentary employment evidence from various sources suggests a rapid rise of employment in the eastern oil fields and a reduction in the number of workers in some of the western oil fields. The Yumen oil field, China's first natural oil base, reportedly hired 14,000 workers in 1957, nearly 20 percent of the industry's total for that year.[10] In the 1958-59 period the number of employees in the Yumen oil field was officially proclaimed as being "several tens of thousands."[11] The dispatch of two-thirds of the Yumen work force to support other new oil fields in the 1960s provides an outstanding example of this eastward thrust of employment. By 1973 the number of workers and employees in Yumen was reportedly 17,000.[12] This figure is greater than 1957 but certainly lower than the "several tens of thousands" officially referred to for 1958-59. In 1959, at the zenith of exploration in Tsaidam, there were reportedly 180,000 workers.[13] However, this number is believed to have fallen below 100,000 in recent years. This seeming flux of employment in a given field is also evidenced in the newly established oil fields. The Tach'ing oil field had 120,000 oil workers in mid-1974.[14] Subsequently some of its workers were sent southward to support Takang and other new fields. Presumably some of this flux represents a flow of skilled personnel who specialize in certain phases.

While employment in individual oil fields fluctuated, total employment for the petroleum industry has registered a continuous rise throughout the entire period. Table 7.4 summarizes the estimated employment in the petroleum industry, based on the limited data available, and relates it to total industrial employment.

From Table 7.4 it can be seen that employment in the petroleum industry increased 100 times between 1949 and 1974, while crude oil output rose 500 times during the same period. In 1974 petroleum accounted for 8.2 percent of the gross value of industrial output, but its relative share of employment was only 2.7 percent. These figures imply a steady rise in labor productivity for the petroleum industry. This rise is the combined result of three factors: the high capital-labor ratio, improvements in technology, and the high prices associated with petroleum products.

Compared with other industries in China, petroleum received an above-average fixed investment per productive worker. During the 1952-55 period, for which official statistics are available, among ten major industries only electric power received a higher fixed

TABLE 7.4

Estimated Employment of All Workers and of Employees
in the Petroleum Industry in Selected Years

Year	Employment (in thousands) Petroleum Industry	All Industry	Petroleum Employment as Percentage of All Industry
1949	12	3,059	0.4
1952	22	5,263	0.4
1953	31	6,121	0.5
1954	39	6,370	0.6
1955	48	6,121	0.8
1956	56	7,480	0.7
1957	67	7,907	0.8
1958	124	22,984	0.6
1974	1,162	43,500	2.7

Sources: (1) Figures for 1949-58 are from Philip Emerson, Nonagricultural Employment in Mainland China, 1949-1958 (Washington, D.C.: Government Printing Office, 1955), p. 43, Table 15.

(2) The 1974 petroleum employment estimate is based on the following information and assumptions. (a) There were an estimated 612,000 workers and employees in the oil extraction industry, including 120,000 at Tach'ing, 100,000 at Shengli, 60,000 at Takang, 17,000 at Yumen, 60,000 at Karamai, 50,000 at Tsaidam, 30,000 at Central Szechwan, 40,000 at Fushun, 35,000 at Maoming, and about 100,000 at other oil fields. (b) In the refining system, employment in 1974 was estimated at 300,000. For instance, the Peking Refinery employed more than 10,000 workers in 1972, New York Times, August 2, 1972. Since there are now 20 refineries comparable to the Peking Refinery, plus several dozen small refineries, the total employment in the refining industry should approach 300,000. (c) In the oil transportation system, an estimated 250,000 workers are employed.

(3) Between 1957 and 1974 the output value of industry rose 4.5 times. (See Table 7.2.) This rate is also applied to employment. Since Chinese industries were more capital-intensive in 1957 and more labor-intensive in recent years, even allowing for some improvement in labor productivity, the output-labor ratio would not be significantly altered during the 1957-74 period.

investment per production worker than petroleum. (See Table 7.5.)
Fixed assets per production worker for the petroleum industry, when
converted into U.S. dollars at the prevailing exchange rates, amounted
to more than $10,000 per worker. This sum is much greater than that
for the average industrial worker. The high capital-labor ratio helps
to explain the high labor productivity.[15]

TABLE 7.5

Fixed Assets for Industrial Production per
Productive Worker in Petroleum and Other
Basic Industries, 1952-55
(in 1952 yuan)

Industry	1952	1953	1954	1955
Overall average	5,656	5,273	6,072	6,835
Petroleum	24,945	NA	NA	27,785
Electric power	NA	51,197	58,828	58,196
Coal mining	5,029	NA	NA	5,417
Iron and steel	9,251	9,241	11,662	13,302
Metal processing	4,750	5,029	5,528	6,035
Chemical	8,120	9,066	9,867	11,114
Building materials	2,431	2,291	2,531	3,641
Food	NA	3,373	3,312	3,566
Paper	9,528	8,923	9,856	10,307
Textiles	4,086	4,943	5,125	5,107

NA = Not available.

Note: The term "productive workers" refers to all workers
and employees who directly participate in productive activities.
Specifically, they are workers, apprentices, trainees, engineers,
technicians, and managerial personnel. In 1952, of the 22,000 workers and employees in the petroleum industry, only 13,250, or about
60 percent, were counted as productive workers. (Kang Chao, Rate
and Pattern of Industrial Growth in China [Ann Arbor: University of
Michigan Press, 1965], p. 38.)

Source: Ch'en Nai-ruenn, Chinese Economic Statistics
(Chicago: Aldine Publishing Company, 1967), pp. 150-51.

Improvements in technology also contribute to the rise in labor productivity. In the initial period most of the oil workers fell in the relatively unskilled category. Official statistics indicate that only 10 percent of all the workers in the petroleum industry were classified as skilled labor.[16] The industry appears to have had relatively few specialists. Strenuous efforts have been made to build a sizable technical force: by 1965, but prior to the Cultural Revolution, there were 5 petroleum institute, 3 geological institutes, 22 universities with geological departments, and 25 secondary schools specializing in geology. The Peking Petroleum Institute, for example, graduated nearly 7,000 students between 1953 and 1964, half of whom specialized in prospecting, mineral geophysics, oil-well engineering, and oil mining. The remainder specialized in oil and natural gas, automation for petroleum processing, basic organic synthesis, and mechanical engineering for mining and refining.[17]

Additionally, training centers were set up in almost all of the major oil fields, and Yumen was assigned the special function of training most of the technicians and skilled workers for other fields. Since the early 1950s the oil fields and training centers have trained more than 100,000 workers and technicians.[18] At a more advanced level, research and training are also conducted at the Petroleum Research Institute of the Chinese Academy of Sciences, the Peking Petroleum Refining Research Institute, the Fushun Research Institute, the Lanchow Research Institute, and the Research Academy of the Ministry of Petroleum Industry. Among the more than 38,000 professional men, technicians, and skilled workers trained in the USSR between 1950 and 1960, many specialized in petroleum engineering.[19] By the end of 1957 official sources claimed that total research personnel in the petroleum industry had increased eight times since 1952.[20]

This growth in technical manpower has proved conducive to technological improvement. According to official reports, the country's average monthly footage per drill in 1972 was 21 percent higher than that in 1965. In 1972 the Number 1205 drilling team at Tach'ing reportedly established a new record by drilling a medium-deep well with one drill in a single day. Another stellar performer, the Number 3252 drilling team, achieved an annual drilling footage of 151,420 meters in a year. All of these officially claimed drilling speeds compare favorably with existing world records.[21]

In recent years, new techniques have been adopted for extracting the oil and injecting water into separated oil strata, in order to keep the pressure stable. Chinese officials claim that by using this new technique the output of high-yield oil wells can be sustained and that the technique has turned some low-yield oil wells into high-yield wells.[22]

Technological improvements have played a major role in bettering the financial position of the industry. During the 1950s the industry yielded a very low rate of return on investment, but the record since the early 1960s reflects a high rate of return. During the First Five-Year Plan period, the profit accumulation accounted for less than half the state investment in the petroleum industry.[23] During 1960-74, however, profit accumulation has far exceeded state investment. In Tach'ing, the taxes and profit surrendered to the state from 1960 to 1974 were reported to have been 11 times the state investment.[24] The profits accrued by the Lanchow Refinery between 1960 and 1966 were reported to be twice the size of the state investment.[25]

The contributions of the petroleum industry to industrial output, productivity, and capital accumulation have been distorted by the pricing system used in China. Chinese prices are based on an average-cost-plus method whereby the price of a final product is composed of the average cost of the input of all the enterprises involved in producing the final product, a profit margin calculated on the basis of costs, and a tax proportionate to the sale value.[26]

In the 1950-52 period the production cost of crude oil was extremely high and the consequent price for crude and oil products was also high. In 1952 the price for one ton of crude oil was 160 yuan, compared with 12.5 yuan for coal.[27] One official source gave a comparison of the prices of diesel oil and gasoline in China, the USSR, and the United States in 1957 as shown in Table 7.6.

TABLE 7.6

Wholesale Prices of Diesel Oil and Gasoline in
China, the USSR, and the United States, 1957
(per ton)

	China	USSR	United States
Diesel oil	220 yuan	164.4 yuan (168.6 old rubles)[a]	78.72 yuan ($30)[b]
Gasoline	640 yuan	204.7 yuan (211 old rubles)[a]	89.22 yuan ($34)[b]

[a]One old ruble equals 0.975 yuan.
[b]One dollar equals 2.624 yuan.

Source: Shih-yu Kung-yeh T'ung-hsun, no. 9 (1957), p. 2.

These figures make it clear that the wholesale prices of Chinese petroleum products were much higher relative to those of other countries. The Chinese price of diesel oil was 40 percent higher than that of the USSR and approximately three times greater than that of the United States. In the case of gasoline, the Chinese price was three times that of the Soviet Union and seven times that of the United States.

In 1969, after Tach'ing was put into large-scale operation, the retail price of gasoline in China was still 3.23 times that of Japan, and that of kerosene was 6.33 times greater than the Japanese price.[28] Compared with the going price of coal in China, petroleum prices were exorbitantly high. According to official statistics, the average price of coal at the mine head in 1956 was 13.018 yuan per ton. Of this price, .948 and 1.195 yuan represented taxes and profit respectively, and the balance of 10.874 yuan represented production costs.[29] The price of crude oil was 160 yuan per ton, about 12 times that of coal. Of the total crude oil price, 30 percent was accounted for by profits and taxes.[30] The relatively high profit margins and taxes assigned to petroleum products, as well as the high cost of production, has tended to give an upward bias to the gross value of petroleum. This upward bias also gives an upward thrust to the value of gross output and labor productivity.

THE IMPORT-SUBSTITUTION EFFECT

Another significant contribution of the Chinese petroleum industry to the national economy is its import-substitution effect, which tends to stimulate greater self-sufficiency for China's crude oil supply. Import-substitution is the difference between the actual imports at the end of a given time period and what they would have been, had the proportion of imports to the total petroleum supply remained constant throughout the period. In short, by applying the import ratio (imports/total supply) of 1950 to the aggregate supply of crude oil and petroleum products for the 1950-65 period and then comparing these results with the actual imports of crude and petroleum products for the same period, one can measure the import-substitution effect.

Methodologically, the change in imports from a base year to a current year can be expressed as

$$dm = m_1 S_1 - m_0 S_0$$

where m represents the import-proportion of petroleum supplies and S denotes total petroleum supplies. The subscripts 0 and 1 refer to the base year and the current year respectively. This import change can be divided into two components

$$dm = S_1 (m_1 - m_0) + m_0 (S_1 - S_0)$$

where the first term $S_1 (m_1 - m_0)$ represents import-substitution effects and the second term $m_0 (S_1 - S_0)$ denotes the expansion of imports due to the increased domestic demand.[31]

In the 1950-61 period, China relied heavily on petroleum supplies from the USSR and Romania. Between 1959 and 1961 China imported more than 3 million tons of crude and petroleum products annually. (See Table 7.7.) Imports of crude and petroleum products during this time accounted for 7 to 11 percent of China's total imports. Imports of crude oil and petroleum products dropped sharply after 1963, partly as a consequence of an increase in domestic production and partly because of stagnating demand. By 1965 China had, for the most part, achieved self-sufficiency in crude oil supply. (See Table 7.8.)

By using data from Table 7.8, the import substitution and demand expansion of petroleum products in various periods can be calculated as in Table 7.9. The results indicate that imports of crude oil and petroleum products rose from 281,000 tons in 1950 to 1,528,000 tons in 1955, an increase of 1,301,000 tons. The change in imports came from two components, the rise of the import ratio from .58 to .62 and the expansion of demand from 481,000 to 1,693,000 tons. In other words, there was no import substitution during this period.

In the following five years (1960-65), despite an increase in demand of 5.4 million tons, domestic supply rose by 6.7 million tons. Consequently, the import-substitution effect outweighed the demand expansion. Thus one can readily perceive why China was in a position to export oil products to North Vietnam and North Korea after 1965. Although the exact quantity of exports is unknown, it is probable that the values of exports and imports of petroleum products during the 1965-72 period were in balance; thus, China could be said to be self-sufficient in petroleum.

In 1973 China began to export crude oil and petroleum products to the non-Communist world. One million tons of Tach'ing crude oil were exported to Japan in 1973, earning $32.6 million. Exports expanded in 1974, with 4 million tons of crude shipped to Japan; 125,000 tons of light diesel oil to Thailand; and another 300,000 tons of petroleum products to Hong Kong. Oil revenue in 1974 exceeded $400 million. Available evidence suggests that China may export 7-10 million tons of Tach'ing crude to Japan and 750,000 tons to the Philippines in 1975. Total oil receipts may exceed $1 billion during 1975.

The emergence of China as a net exporter of crude oil and petroleum products at a time when international oil prices rose 400 percent in two years has had a great impact on the Chinese balance of payments. Although oil revenue represented only 6 percent of the

TABLE 7.7

Imports of Crude Oil and Refined Products, 1950–68
(in thousands of tons)

Year	From USSR			Value (in millions of U.S. dollars)	From Other Countries	Total
	Crude	Petroleum Products	Total			
1950	--	281	281	11.2	--	281
1951	--	729	729	NA	--	729
1952	--	608	608	32.5	--	608
1953	--	834	834	44.5	--	834
1954	--	904	904	48.5	--	904
1955	378	1,204	1,582	79.0	--	1,582
1956	397	1,335	1,732	86.0	--	1,732
1957	380	1,422	1,803	90.5	--	1,803
1958	672	1,835	2,507	92.3	--	2,507
1959	568	2,412	3,048	117.5	246	3,294
1960	0	2,372	2,940	113.0	333	3,273
1961	0	2,903	2,903	120.7	114	3,017
1962	0	1,856	1,856	80.5	93	1,949
1963	0	1,408	1,408	60.7	376	1,784
1964	0	505	505	--	241	746
1965	0	38	38	--	237	275
1966	0	40	40	--	--	40
1967	0	7	7	--	--	7
1968	0	1	1	--	--	1

NA = Not available.

Sources: (1) 1952–60 figures are from K. C. Yeh, Communist China's Petroleum Situation (Santa Monica, Calif.: Rand Corporation, 1962), p. 43; (2) figures for 1950–51 and 1961–68 are from Kambara Tatsu, "The Petroleum Industry in China," China Quarterly, December 1974, p. 703; (3) value figures for 1952–60 are from Yeh, op. cit., which gives them in old rubles; the ruble figures are converted into dollars at 1 dollar equals 4 old rubles; (4) the value figure for 1950 is from China News Analysis (Hong Kong), no. 220 (March 14, 1958), p. 6; (5) the value figures for 1961–63 are from Alexander Eckstein, Communist China's Economic Growth and Foreign Trade (New York: McGraw-Hill, 1965), pp. 106–107.

TABLE 7.8

Imports and Total Supply of Crude and Petroleum Products, 1950-65

Year	Domestic Production (in thousands of metric tons)	Crude and Products Imported (in thousands of metric tons)	Import Ratio (in percent)	Self-Sufficiency Rate (in percent)	Total Oil Supply (in thousands of metric tons)
1950	200	281	58	42	481
1951	305	729	73	27	1,134
1952	436	608	58	42	1,044
1953	622	834	57	43	1,456
1954	789	904	53	47	1,693
1955	966	1,582	62	38	2,548
1956	1,163	1,732	60	40	2,895
1957	1,458	1,803	55	45	3,261
1958	2,264	2,507	53	47	4,771
1959	3,700	3,294	47	53	6,994
1960	5,200	3,273	39	61	8,473
1961	6,000	3,017	34	66	9,017
1962	6,700	1,949	23	77	8,649
1963	7,500	1,784	20	80	9,284
1964	8,500	746	9	91	9,246
1965	11,000	275	2	98	11,275

Sources and Notes: Domestic output figures are from Table 7.1 (sources are given in Tables 2.1, 2.2, and 2.4); import figures are from Table 7.7; the import ratio is computed by dividing imports by oil supply; the self-sufficiency rate is domestic production divided by oil supply.

value of China's exports in 1974 and an estimated 7 percent in 1975, the petroleum industry accounted for more than 15 percent of China's balance of payments during these two years, if one includes the impact of the import-substitution effect.

TABLE 7.9

Effect of Import Substitution and Expansion in Demand
on Imports of Crude Oil and Petroleum Products
During Selected Periods, 1950-65
(in thousands of metric tons)

Period	Import Substitution	Expansion in Demand	Total Change in Imports
1950-55	93	1,208	1,301
1955-60	-1,988	3,679	1,691
1960-65	-4,080	1,082	2,998
1955-65	-6,715	5,408	-1,307

Note: Figures in this table are calculated from the equation

$$dm = S_1 (m_1 - m_0) + m_0 (S_1 - S_0)$$

For example, the 1950-55 figures are derived in the following way:

Import Substitution = 2,548 (0.6209 - 0.5842)
= 93
Expansion in Demand = 0.5842 (2,548 - 481)
= 1,208
Total Change in Imports = 93 + 1,208
= 1,301

Source: Figures are derived from Table 7.8 (sources are from Tables 2.1, 2.2, 2.4, and 7.7); the equation is explained in the third section of Chapter 7 of the text.

THE FORWARD AND BACKWARD LINKAGE EFFECTS

Less direct, yet equally important to Chinese industrialization, are the forward linkage effects generated by crude oil and natural gas as raw materials to the secondary petrochemical industry.

Industrial development fostered by the use of oil and gas as raw materials has shown phenomenal growth in many of the developed

countries of the world. In 1960 about 32 percent of the total chemical production in the United States came from petroleum-based raw
materials. The percentage rose to 38 in 1965 and 41 percent by
1970. In the United Kingdom the percentage was 47 in 1959 and 60
by 1962.[32] The petroleum-based raw materials consist of organic
chemicals mainly composed of carbon and hydrogen, and generally
oxygen, but sometimes chlorine, sulphur, and nitrogen. There are
also petroleum-based inorganic chemicals, which are largely carbon
and carbon disulphide. The petrochemical industry produces tens of
thousands of products, the development of which constitutes a major
yardstick in measuring progress toward industrialization.

   Although lagging behind the petroleum industry, China's petrochemical industry is gestating in an environment favorable to rapid
growth. Indeed, the expansion of crude oil production and refining
since the early 1960s has resulted in the mushrooming of gas-
consuming industries contiguous to the oil-producing areas. The
government's intention to control and accelerate the development of
the petrochemical industry is evidenced by its action to merge the
coal, petroleum, and chemical ministries into a single Ministry of
Fuel and Chemicals in 1969.[33] More recently, the Ministry of Fuel
and Chemicals was reorganized into two separate ministries: one
specializing in coal industry, the other specializing in petroleum
and chemical industries.

   During the decade between 1965 and 1975, four major petrochemical complexes were constructed or under construction in areas
proximate to the major oil fields. In Peking, a modern petrochemical complex, the Peking General Petrochemical plant, was constructed at Feng-huang-ling, 60 kilometers southwest of Peking.
When construction was begun in 1968 the complex possessed three
major refining installations and processed 2.5 million tons of crude
oil annually; by 1975 its processing capacity exceeded 4 million tons.
The complex, when completed, will ultimately consist of separate
units producing gasoline, kerosene, diesel oil, lubricants, synthetic
rubber, polystyrene, acetones, and a wide variety of chemicals. A
list of plants constructed near the Peking oil refinery site would include the following: a 1,3-cis polybutadiene plant of 3,000 tons per
year, a polypropylene plant of 4,000 tons per year, a caprolactone
plant of 3,000 tons per year, a styrene and polystyrene plant of 4,000
tons per year, and a detergent plant of 1,000 tons per year.[34] The
completion of these plants will make Peking a major petrochemical
center in China. The petrochemical industry located in this area
will utilize crude oil produced in Takang, Tach'ing, and Shengli.

   The second major petrochemical complex under construction
is located in Manchuria, 31 miles south of Shenyang. Plant and
equipment outlays totaled $282 million (1.2 billion francs) and were

supplied by a French consortium led by Technip and Speichem. The entire complex encompasses 17 plants. (See the third section of Chapter 6.) The completion of this complex will probably make Southern Manchuria the leading center of the Chinese petrochemical industry. Nearly contiguous with the refining center of Chin-chou and the newly established refinery at Anshan, this complex will use the crude oil and shale-oil output of Tach'ing, Fushun, and Chin-chou as its source of supply.

The third major petrochemical complex is located at Lanchow, the major industrial center of the northwest. Surrounding the modern Lanchow Refinery (annual capacity 5 million tons) are a large number of petrochemical plants, some already constructed and some under construction. These plants include a high-density polyethylene plant as well as a polypropylene plant. There are now more than 50 plants producing synthetic fabric, synthetic rubber, and plastics in this complex. In 1972 Lanchow reportedly turned out ten times as much plastic as it did in the six years prior to 1965 and 2.2 times as much synthetic fiber and synthetic rubber as in 1965.[35]

The fourth petrochemical complex is located at Hangchow in Chekiang province. Recognized for its textile and silk manufacturing, Hangchow is becoming the leading center for synthetic fabric in China. In 1971 a new refinery with an annual processing capacity of 1 million tons was constructed at Hangchow. During 1972-73, China contracted with Japan for four major petrochemical plants for a price of $151 million. These new plants include (1) a synthetic fiber plant to be completed in 1975, with an annual capacity for producing 30,000 tons of ethylene and 45,000 tons of butadiene; (2) a synthetic ammonia plant to be completed in 1975, with an annual capacity of 330,000 tons of ammonia and 528,000 tons of urea; (3) a high-pressure, low-density polyethylene plant with an annual capacity of 60,000 tons; (4) a high-pressure polyethylene plant with an annual capacity of 180,000 tons.[36] The completion of these four new projects will make Hangchow one of the most important petrochemical centers south of the Yangtze River.

Three petrochemical complexes on a significantly smaller scale have been erected at Hsuchow in North Kiangsu and at Tsupo in Central Shantung, both being close to the Shengli oil field. The third one is being developed at Luchow, southwest of the Nanch'ung oil field in Szechwan.

The Hsuchow petrochemical industry includes a calcium carbide plant, a synthetic rubber plant, a detergent plant, a gas manufacturing plant, and several fertilizer and agricultural chemicals plants. In 1972 the output of this area was officially reported as having increased six times over its output in 1958, at a value of more than 100 million yuan a year.[37]

At Tsupo, in central Shantung, a new refinery with an annual processing capacity of 1 million tons was completed in 1971. A number of petrochemical plants are reportedly under construction to utilize its refining products.[38]

Since 1966 the Luchow chemical complex, located between Chungking and Tsukung in one of the major natural-gas fields in Szechwan, has added several modern petrochemical plants, including a high-pressure polyethylene plant for plastics. The polyethylene plant, which is equipped with British machinery, has an annual capacity of 24,000 tons.

Parallel to developments in the new centers, vigorous expansion of petrochemical facilities has been undertaken at such old industrial bases as Shanghai, Nanking, Tientsin, and Kirin. By 1974 a young petrochemical industry capable of producing synthetic fiber, resins and plastics, rubber, agricultural chemicals, detergents, and a wide range of industrial chemicals had emerged in China.

In the field of synthetic fiber, more than a dozen modern plants were added at Peking, Shanghai, Ch'ang-chun, Hangchow, Lanchow, Luchow, Tsupo, Chungking, and Tientsin. Most of their equipment was made in Japan and Western Europe. According to one U.S. expert who visited China in 1972, China's synthetic-fiber producing capacity of 1972 can be estimated at 25,000 tons of polyamides (nylon-type fibers), 25,000 tons of polyester (Dacron), and 15,000 tons of polyacrylics (Orlon).[39]

In 1972-74 China contracted with Japan for five new synthetic fiber plants, most of which will be constructed at Hangchow. The completion of these new facilities will increase China's synthetic-fiber producing capacity to 300,000 tons a year.[40]

In the field of synthetic resins and plastics, no definite information is available. According to one study made in Taiwan, there were 195 plants of various sizes engaged in plastics production in 1970, most of which, however, were small in scale and primitive.[41] Two major plants were built in Peking, the Peking Vinyl Chloride Plant, specializing in polyvinyl chloride (PVC), with a capacity of 30,000 tons a year; the Peking Petrochemical Plant, which specializes in low-pressure polyethylene and polystyrene.[42] Plastics raw-material plants were also founded at Chinhsi in Liaoning, producing vinyl chloride, vinyl resins, and organic glass; at Lanchow, producing acrylic resins; and at Luchow, producing high-pressure polyethylene. In 1957 the output of plastics was officially given as 10,000 tons a year.[43] No overall statistics are available for the period after 1957. One Japanese expert estimated that the output of plastics in 1966 was 60,000 to 70,000 tons.[44] Scattered data in several producing centers indicate a continued rise in production. In Peking, where most of the modern plants are located, the output of plastics raw materials reportedly registered a threefold increase between 1965 and 1972. [45] In Lanchow, an acrylic resin plant supplied by the United Kingdom in 1968 produced 50,000 tons a year. One Taiwan expert estimated the 1972 output for the whole plastics industry at 200,000 tons a year. [46] It may have reached 300,000 tons in 1974.

Chinese synthetic rubber manufacturing is still relatively primitive; only three or four factories appear to have any significant potential. A plant within the Kirin Chemical Company, the most important chemical complex in China in 1974, has been producing the styrene-butadiene-type synthetic rubber at a rate of 30,000 tons a year. [47] A new plant built in 1969 at Chinhsi in Liaoning produces butyl rubber, and the Peking Chemical Engineering Plant produces polybutylene at a rate of 15,000 tons a year. The total amount of synthetic rubber produced in China in 1974 would appear to have been in the range of 60,000 to 70,000 tons.

The synthetic detergent industry is also in its initial stage of development. Several plants were constructed in Tientsin and Shanghai in 1964, using alkyl aryl sulphonates and alkyl sulphates as raw materials. Their scale of operation remains relatively small. One major plant was reportedly built at Fushun in 1966 utilizing oil shale as raw material. This plant produced 78,000 tons of aromatic hydrocarbon a year. Total synthetic detergent production in 1973 was estimated by a Taiwan expert at only 50,000 tons. [48] Given the capacity of the Fushun plant, however, output may have reached 100,000 tons in 1974.

In contrast to detergents, agricultural chemicals have had rapid development since the early 1960s. China can now produce the major chemicals used to destroy or ward off plant diseases. DDT and BHC (benzene hexachloride) rank among the more significant products. China began producing DDT and BHC in 1953 and has substantially accelerated their production since 1962. Today almost every province in China possesses several plants producing diverse basic agricultural chemicals, and total output has been reported as 91,000 tons in 1958 and 137,000 tons in 1959. [49] Between 1965 and 1973 the output of agricultural chemicals has reportedly increased "several times." [50] China's output of agricultural chemicals may well have exceeded 1 million tons in 1974.

Although it is far from exhausting the whole gamut of petrochemicals, one can still be aware from the several branches surveyed that China's petrochemical industry is in position and ready to take off. In the past the development of this new industry has been hampered by the use of outmoded equipment and obsolete processes; [51] this difficulty will be greatly mitigated with the completion of the more than 15 new plants ordered from Japan and Europe during 1973-74.

The forward-linkage effect generated by the petroleum industry on the chemical industry can be roughly gauged by comparing the relative growth of the Chinese chemical industry during various periods. As shown in Table 7.4, the relative share of the chemical industry in the total output of industry remained quite steady between 1957 and 1964, varying between 5.4 and 6 percent. However, the rapid development of petrochemicals in the 1965-74 period provided additional thrust to the chemical industry's share of total industrial output. The chemical industry's share relative to total output was 8.1 percent in 1965, 9.1 in 1972, and well over 10 percent during 1974.

The available evidence suggests that this rapid increase stemmed from the growth of petrochemicals. In 1957, prior to the expansion of the petroleum industry, the combined output value of the petroleum and chemical industries accounted for only 6 percent of China's industrial gross output value. As a consequence of the phenomenal growth in petroleum and petrochemicals, the combined output value of the petroleum and chemical industries in 1974 represents almost one-fifth of the total industrial output value. Given the surge of growth reflected in petroleum output and the concomitant expansion of petrochemical facilities, the future production possibilities for these two closely related industrial branches appears very promising.

The derived-demand or backward-linkage effects generated by the petroleum industry to other industries are best exemplified by the rapid development of petroequipment manufacturing, tanker building, and special steel production.

As noted in Chapter 5, prior to 1954, with the exception of a few factories making maintenance equipment and parts, there were virtually no specialized plants producing petroequipment in China. Most of the equipment needed in the 1949-57 period was supplied by the Soviet Union.

The rift with the Soviet Union and the opening of Tach'ing in 1960 provided a major impetus for the development of a petroleum equipment industry. The output of petroequipment in 1963 was officially reported as 60 percent higher than that in 1962 and to have doubled in 1964.[52] Twelve heavy machinery plants in Shanghai, Shenyang, and Harbin were converted to producing refining equipment. By 1964 the petroequipment industry in China could produce coke towers for decomposing heavy oil into light petroleum, extraction columns, heat exchangers, and various oil pumps. Until 1960 the petroleum and chemical equipment industry was only a subsection of the heavy mining-machinery industry; since 1961 it has become an independent division of China's machine-building industry. During the decade between 1965 and 1974 this division outperformed the others in the machine-building industry in output value, and as a result, petroleum equipment currently ranks as a major division among the 13 divisions of China's machine-building industry.[53]

The year 1965 represents an important benchmark for this division, in that self-sufficiency was basically attained in terms of producing most of the petroleum equipment for onshore exploration.[54] The industry is now capable of producing a wide range of products that satisfies approximately 80 percent of the requirements for onshore prospecting and exploration. In the field of refining, China can manufacture an integrated set of refining installations for a designed capacity of 2.5 million tons a year, as exemplified by the newly constructed Peking General Petrochemical Plant. In the field of transport equipment, China has built ocean and coastal tankers up to 25,000 tons and manufactured steel pipe with diameters up to 30 inches (75 centimeters). Recent reports indicate that the Chinese have designed and built their first floating drilling vessel for offshore exploration. The Kantan (Prospector) Number 1 has successfully drilled a well for deep-water oil prospects in the south part of the Yellow Sea.[55] This provides the first tangible evidence that the Chinese are tooling up to produce offshore drilling equipment.

The steel and nonferrous metals industries have also been stimulated by the petroleum industry. Before 1960 China produced only limited quantities of annealed steel, which was used for making high pressure containers. Most of the alloy steel, seamless steel pipes, and stainless steel plates had to be imported. After 1964, to meet the pressing demand for petroequipment, the Chinese metallurgical industry began to produce low-temperature resisting steel plates for use in the manufacturing of oceangoing tankers. They also began to produce composite stainless steel plates for the manufacture of nitrogen fertilizer equipment and high-pressure hydrogen-resisting steel pipes for use in manufacturing oil pipes.

In 1971-74 China's metallurgical industry trial-produced more than 100 kinds of new steel. Many of these have had a favorable impact on petroleum equipment manufacturing.

APPRAISAL OF CONTRIBUTIONS

So far, discussion has centered on four aspects of the petroleum industry's contribution to the national economy. More detailed analysis requires cognizance of the influence of other related factors.

As previously noted, the gross output value of the petroleum industry may be inflated because of the high price placed on petroleum output in the early years of development. Close approximations of the contributions of this industry can be measured by the

value-added approach. According to the study by T. C. Liu and
K. C. Yeh, of China's national income for the 1952-57 period, the
rate of value-added (net value added/gross output value) for crude
oil was lowest among the six identified mineral products (see Table
7.10), being 57 percent in 1952 and 60 percent in 1957, as opposed
to 67 percent and 68 percent for all mineral products in the corresponding years.[56]

TABLE 7.10

Gross Output and Value-Added for Six Mineral
Products, 1952 and 1957

	Gross Output Value (in billions of 1952 yuan)		Net Value-added (in billions of 1952 yuan)		Rate of Value-added (in percent)	
	1952	1957	1952	1957	1952	1957
Coal	0.79	1.54	0.58	1.15	73	74
Crude oil	0.07	0.23	0.04	0.14	57	60
Iron ore	0.08	0.39	0.07	0.30	87	70
Manganese ore	0.01	0.03	0.01	0.03	100	100
Limestone	0.10	0.22	0.09	0.20	90	90
Salt	0.61	1.12	0.54	0.90	88	80
Total	2.17	4.45	1.47	3.07	67	68

Source: Ta-chung Liu and Kung-chia Yeh, The Economy of
the Chinese Mainland, National Income and Economic Development
(Princeton: Princeton University Press, 1965), p. 573, Table H-2.

The Liu-Yeh study did not provide the value-added for those
petroleum products that are actually more important than the crude
oil itself; as a consequence, they failed to present the value-added
for the entire petroleum industry. According to one official analysis
of the production costs of petroleum products, 68 percent of the outlays were for raw material and power consumptions in 1956-57.[57]
From these data one can infer that the rate of value-added for petroleum products was approximately 32 percent. In the known 1956
gross output value of 650 million yuan, 186 million yuan was derived

from crude oil output and 464 million yuan was derived from petroleum products.* Using the Liu-Yeh value-added rate for crude oil and the official value-added rate for petroleum products, the value-added for the petroleum industry can be derived for the year 1956 as 260 million yuan (186 x .6 + 464 x .32 = 260) and the rate of value-added for the petroleum industry can be calculated as 40 percent (260 million ÷ 650 million = 40 percent).

The derived value-added rate of the petroleum industry when compared with other producer industries (see Table 7.11) ranked very low. This further suggests that the real contribution of the petroleum industry to the GNP is somewhat smaller than its relative share of the gross output value.

Although employment in the petroleum industry grew at a much slower rate than its gross output value, the Chinese petroleum industry shows signs of overemployment when compared with other countries. Many Japanese and Western experts who have visited production facilities in China in recent years have testified to the over-staffing of the petroleum industry. Herman F. Mark, a leading American polymer chemist who visited many laboratories, two nylon plants, and one oil-refining complex, found employment in the Peking General Petrochemical Plant to exceed 10,000 people, whereas a comparable American plant would be operated by around 800 employees.[58]

Similar observations were also made by a well-known Japanese expert, Genko Uchida, who found a vinyl chloride factory in Peking that employed 2,700 workers in its polymerization division, as contrasted with only 150 to 200 employees in a comparable-sized plant in Japan.[59]

Based on these assessments, one may conclude that the Chinese petroleum and petrochemical industries are grossly overstaffed, which helps to explain the undue expansion of employment.

On the other hand, the import-substitution effect does not fully reflect the contribution of the petroleum industry. In the 1950-70 period, capital investment in China was limited by her capacity for importing machinery and equipment from abroad, which in turn was conditioned by her capacity for export. The autarkic policy pursued by China in the 1960s stemmed primarily from her inability to export

---

*These figures are derived in the following way: (1) the gross output value for the entire petroleum industry was given as 650 million yuan; (2) the gross output value of crude oil is equal to the output of crude oil (1,163,000 tons) times the price per ton (160 yuan), which is 186 million yuan; (3) the difference between the gross output value for the industry and the gross output value of crude oil is the gross output value of petroleum products, which equals 464 million yuan.

TABLE 7.11

Rate of Value-Added of Petroleum Industry Compared
with Other Seven Major Industries, 1957
(in percent)

Industry	Rate of Value-added
Machinery	0.460
Ferrous metallurgy	0.822
Building materials	0.604
Chemicals	0.370
Petroleum*	0.400
Crude oil	0.600
Petroleum products	0.320
Power	0.544
Coal	0.646
Timber	0.970

*Although the value-added rate for petroleum is based on 1956
figures, no significant change is for 1957.

Sources: Figures for petroleum, crude oil, and petroleum
products are from Ta-chung Liu and Kung-chia Yeh, The Economy
of the Chinese Mainland, National Income and Economic Development (Princeton: Princeton University Press, 1965), p. 573, Table
H-2; figures for other industries are from Thomas G. Rawski's
paper, "Measuring China's Industrial Performance 1949-72."

in that period. During recent years, as China began to export crude
oil in quantity, this autarkic policy has been displaced. To speed
the modernization of industry, China has been purchasing enormous
amounts of Western machinery and technology. Imports of machinery and equipment totaled $395 million in 1970, $505 million in 1971,
$520 million in 1972, and $855 million in 1973--more than doubling
in four years. In 1973 China signed contracts for whole plants (turn-
key projects) and other machinery and equipment valued at almost
$2.5 billion. Most of the $1.2 billion in whole plants purchased will
produce chemical fertilizer, synthetic fibers, and petroleum-based
plastics. Other purchases include transport equipment, machines
for the mining and petroleum industries, and dredgers for port improvement. In the first half of 1974 China purchased additional
plants from companies in Japan and Western Europe valued at $750
million. The largest deal is for a $430 million steel-rolling complex

purchased from West German and Japanese firms.[60] The roseate prospects for increasing crude oil exports have thus tended to speed up China's modernization at a rate beyond that warranted by its initial effect on the balance of payment.

Without detailed input-output analysis, precise evaluation of the linkage effects generated by the petroleum industry is not feasible. However, the impact of the dispersion of the petroleum industry will be profound as China's infant petrochemical industry enters the stage of sustained growth. The linkage effect may not be limited to closely related industries but may also extend to agriculture. One official analyst estimated that the gases generated by a refining plant with an annual processing capacity of 600,000 tons, if fully utilized, can produce 3,000 tons of polyphenyl ethylene, 10,000 tons of alcohol, 6,000 tons of phenol hydroxybenzene, and 3,600 tons of acetone (Propanone). From the phenol hydroxybenzene alone, 5,000 tons of synthetic fiber materials can be manufactured (the equivalent of 35,000 tons of cotton), from which 200 million meters of cloth can be made.[61] Since Chinese refineries now process over 40 million tons of crude a year, the full utilization of its byproducts may enable China to reduce its demand for cotton, thus alleviating the competitive demands of cotton and food grain for scarce arable land.

NOTES

1. Albert O. Hirschman, The Strategy of Economic Development (New Haven: Yale University Press, 1966), pp. 98-119.

2. P. N. Rasmussen, Studies in Intersectoral Relations (Amsterdam: North-Holland Publishing Co., 1956), pp. 140-41.

3. Peter R. Odell, An Economic Geography of Oil (London: G. Bell and Sons, 1963), pp. 183-85.

4. Chao I-wen, Hsin-chung-kuo ti Kung-yeh (Peking: T'ung-chi Ch'u-pan-she, 1957), p. 50.

5. Ibid.

6. Hsueh Shih-shih, no. 1 (1975), p. 6.

7. People's Republic of China, State Statistical Bureau, Ten Great Years (Peking: Foreign Languages Press, 1960), p. 87.

8. Peking Review, no. 4 (January 24, 1975), p. 22.

9. John Philip Emerson, Nonagricultural Employment in Mainland China, 1949-1958 (Washington, D.C.: Government Printing Office, 1965), p. 106.

10. Kung-jen Jih-pao, December 20, 1957.

11. Shih-yu K'an-t'an, no. 16-17 (August 1958), pp. 7-9.

12. China Reconstructs, December 1973, p. 32.

13. Jen-min Jih-pao, February 2, 1959.

14. Yomiuri Shimbun (Tokyo), July 2, 1974.

15. According to a study on labor productivity in Chinese industry by Robert M. Field, a main reason for the rise in labor productivity in the 1952-57 period was the rise in the capital-labor ratio. Alexander Eckstein et al., eds., Economic Trends in Communist China (Chicago: Aldine, 1968), p. 655.

16. K. C. Yeh, Communist China's Petroleum Situation (Santa Monica, Calif.: Rand Corporation, 1962), p. 40.

17. Chu-yuan Cheng, Scientific and Engineering Manpower in Communist China (Washington, D.C.: National Science Foundation, 1966), pp. 273-74.

18. Lieh wen, "Yumen Oil Men in Great Leap Forward," Chung-kuo Hsin-wen, February 26, 1959.

19. Chu-yuan Cheng, op. cit., pp. 196-97.

20. Shih-yu Lien-chih, no. 2 (1959), p. 1.

21. Hua Ching-yuan, "New Achievements in China's Oil Industry," China's Foreign Trade (Peking), no. 1 (1975), pp. 6-7.'

22. Ibid.

23. Shih-yu Kung-yeh T'ung-hsun, no. 3 (1957), p. 2.

24. Peking Review, June 7, 1974, p. 17.

25. Kung-jen Jih-pao, March 25, 1966, p. 2.

26. Ch'en Nai-ruenn, Chinese Economic Statistics (Chicago: Aldine Publishing Company, 1967), p. 81.

27. Akira Doi, "The Production and Price of Petroleum from Chinese Ta-ch'ing Oilfield," Showa Dojin, August 1970, pp. 18-22.

28. Wu Yuan-li, Economic Development and the Use of Energy Resources in Communist China (New York: Praeger Publishers, 1963), p. 139.

29. Ta-chung Liu and Kung-chia Yeh, The Economy of the Chinese Mainland, National Income and Economic Development (Princeton: Princeton University Press, 1965), p. 569.

30. Akira Doi, op. cit.

31. This analytical technique is borrowed from Alfred Maizels; see his Industrial Growth and World Trade (Cambridge: Cambridge University Press, 1963), pp. 150-52.

32. George Sell, The Petroleum Industry (London: Oxford University Press, 1963), p. 245.

33. Chung-kung Nien-pao 1972 [Yearbook of Chinese Communism 1972] (Taipei, 1972), pp. 2-45.

34. C. K. Jen, "My Impressions of the New China and Its Science and Technology," Eastern Horizon 12, no. 4 (1973): 54-55.

35. New China News Agency (Peking), September 5, 1972.

36. U.S.-China Business Review, January-February 1974, p. 36; also Ta-lu Ching-chi Yen-chiu, October 1974, pp. 51-52.

37. Jen-min Jih-pao, May 13, 1973, p. 2.

38. Jen-min Jih-pao, January 18, 1971, p. 2.

39. This figure was supplied by Professor Herman Mark, an authority on high polymer chemistry who visited China in June 1972; quoted from C. K. Jen, op. cit., pp. 54-55.

40. Yiu I-ming, "Sino-Japanese Trade: Current Development and Future Prospects," Ta-lu Ching-chi Yen-chiu, October 1974, pp. 51-52.

41. Hua Chung-shih, "A Survey of Petrochemical Industry in Communist China," Ta-lu Ching-chi Yen-chiu, October 1974, pp. 15-30.

42. China Reconstructs, March 1966, p. 5.

43. Jen-min Jih-pao, September 11, 1957.

44. Genko Uchida, "Technology in China," Scientific American, November 1966, p. 43.

45. Chung-kuo Hsin-wen, January 20, 1973, p. 5.

46. Hua Chung-shih, op. cit., p. 22.

47. Jen-min Jih-pao, May 21, 1963.

48. Hua Chung-shih, op. cit.

49. Ibid.

50. New China News Agency (Peking), September 29, 1973.

51. C. K. Jen, op. cit., p. 55.

52. Chi-hsieh Kung-yeh, no. 20 (1964), pp. 14-15.

53. For details, see Chu-yuan Cheng, The Machine-building Industry in Communist China (Chicago: Aldine-Atherton, 1971), pp. 197-98.

54. Jen-min Jih-pao, January 1, 1966.

55. Peking Review, January 10, 1975, p. 5.

56. Ta-chung Liu and Kung-chia Yeh, op. cit., p. 573, Table H-2.

57. Shih-yu Kung-yeh T'ung-hsun, no. 8 (1957), p. 1.

58. New York Times, August 2, 1972, pp. 47, 52.

59. Genko Uchida, op. cit., p. 43; and Chuo Korun (Tokyo), July 1967.

60. Chu-yuan Cheng, "Communist China's Machine-building Industry: New Trends in Production and Trade," a paper presented to the 25th Annual Meeting of the Association for Asian Studies, March 31, 1973; also Central Intelligence Agency, People's Republic of China: International Trade Handbook, September 1974, p. 5.

61. Shih-yu Lien-chih, no. 6 (1960), p. 1.

8

THE POTENTIAL OF
CHINA'S TRADE IN OIL

The analysis in the preceding chapters leads to the tentative conclusion that China is on the threshold of becoming a significant net exporter of crude oil and petroleum products. Obviously, to sustain recent growth rates, China will have to rely on the importation of new equipment and technology from advanced industrial countries. In this chapter the objectives will be to analyze some of the major factors that will shape China's foreign trade in the immediate future and to forecast through 1985 the possible levels of crude oil exports from China to the United States and Japan in exchange for the petroleum equipment, technology, and other commodities needed by the Chinese economy.

CHINA'S FOREIGN TRADE IN RETROSPECT

As in other large continental economies, China's foreign trade constitutes a relatively small percentage of her GNP, ranging from 3 to 6 percent. Though small in volume, foreign trade nevertheless plays a crucial role in China's economic development, since it provides an important channel for the acquisition of advanced technology and the capital goods that are vital to her industrialization.

In the pre-1949 era, China's foreign trade was Western-oriented. In 1936 the United States accounted for 22 percent of China's exports and 20 percent of her imports. In 1946 the share was as high as 54 percent for China's imports and 48 percent for her exports. Even in 1950, the first year of the People's Republic of China, U.S. trade still constituted 23 percent of China's total trade.[1] Between 1951 and mid-1971 Sino-American trade discontinued, as the United States imposed an embargo on trade with China as a consequence of the latter's participation in the Korean War.

During the 21 years when Sino-American trade was suspended, China's foreign trade relations experienced some dramatic changes. In the first decade of her existence, 1950-59, major trade partners of the People's Republic of China were confined to the Communist-bloc nations. About 40 to 55 percent of China's trade was with the Soviet Union, 20 to 25 percent with other Communist countries, and 25 to 35 percent with the non-Communist world. As a consequence of the deterioration of Sino-Soviet relations, the relative share of China's total trade accounted for by the Soviet Union declined from over 40 percent in 1960 to a negligible 1 percent by 1970. During the same decade, the relative share of the non-Communist world rose from 35 to 80 percent.

Foreign trade volume during these 21 years has fluctuated erratically. Between 1952 and 1959 trade volume rose at an average annual rate of 15 percent. As a result of this increase, the growth rate of foreign trade outpaced the 9 percent annual increase in GNP. In the following decade, 1960-69, these trends were reversed. Foreign trade not only failed to grow, but actually declined absolutely; by 1969 the combined volume of imports and exports was 12 percent below the 1959 peak, while the GNP rose by approximately 40 percent. Thus more emphasis appears to have been placed on greater self-sufficiency and on deriving growth from within the nation itself. [2] (See Table 8.1.)

The "self-reliance" doctrine, which was strongly emphasized in the post-1960 era, has been substantially modified in recent years. After the three years of disruption unleashed by the Cultural Revolution, 1967-69, the Chinese economy experienced a new surge of growth and expansion. Capital investment in industry during 1971-72 increased dramatically. To facilitate this new expansion, China imported an increasing amount of machinery and equipment, including complete plants that cannot currently be manufactured in China, or not for reasonable production costs. Moreover, China's foreign relations underwent a complete reversal following the "ping-pong diplomacy" started in the spring of 1971. Subsequent events, such as the seating of the Chinese delegation in the United Nations, the establishment of diplomatic relations with Japan, and the gradual normalization of relations with the United States, have provided a diplomatic framework amenable to China's vigorous reentry into trading with the non-Communist nations.

As Sino-American relations improved, trade was initiated during the second half of 1971. In the following year U.S. exports to China totaled only $60 million, and U.S. imports from China were a meager $32 million. However, 1973 witnessed a vigorous upturn, and U.S. trade with China totaled $740 million in exports and $60 million in imports. This increase represented a twelvefold gain over

TABLE 8.1

Gross National Product and Foreign Trade, 1952-74

Year	Gross National Product			Foreign Trade		
	Value (in billions of U.S. 1973 dollars)	Index (1952 = 100)	Growth (in percent)	Volume (in billions of U.S. dollars)	Index (1952 = 100)	Growth (in percent)
1952	67	100		1.89	100	
1957	94	140		3.06	161	
1958	113	169	+20	3.76	198	+22
1959	107	159	-6	4.29	226	+14
1960	106	158	-1	3.99	211	-7
1961	82	122	-23	3.02	159	-25
1962	93	138	+13	2.68	141	-12
1963	103	153	+10	2.77	146	+3
1964	117	174	+13	3.22	170	+15
1965	134	200	+14	3.88	205	+20
1966	145	216	+9	4.24	224	+9
1967	141	210	-3	3.90	205	-9
1968	142	211	0	3.76	198	-4
1969	157	234	+10	3.86	204	+2
1970	179	267	+14	4.29	226	+11
1971	190	283	+5	4.72	249	+10
1972	197	294	+3	5.92	313	+25
1973	217	323	+10	9.88	522	+66
1974*	223	332	+2	13.70	724	+39

*1974 figures are preliminary.

Sources: Central Intelligence Agency, Office of Economic Research, People's Republic of China: Handbook of Economic Indicators (Washington, D.C.: Central Intelligence Agency, August 1975), p. 1.

the preceding year and made the United States the second-largest
trade partner of China, second only to Japan.

Sino-American trade in 1974 registered a further increase over
that in 1973. The total volume of trade for 1974 reached $934 mil-
lion, of which $819 million were Chinese imports.[3]

Since the resumption of Sino-American trade, there has been
an increasing differential between U.S. exports to and imports from
China. In 1972 the ratio between U.S. exports and imports was 2:1;
it rose to 12:1 in 1973 and appeared to be approximately 7:1 in 1974.
The fundamental reason for this imbalance is the growing Chinese
demand for U.S. farm products, transportation equipment, and machinery, all products in which the United States has a comparative
advantage. As yet, China has little in the way of exports to offer
the United States. Currently, China finances these trade deficits
with the United States through surpluses earned in Hong Kong and in
Southeast Asia. However, these foreign exchange earnings are
needed to cover not only the trade deficits with the United States but
also those with Canada, Australia, Japan, and Western Europe.
Thus China's ability to import will necessarily be constrained by
her export capability. Given the fourfold increases in world oil
prices, it is readily obvious that China's crude oil and petroleum
products may constitute the needed potential exports. In 1973 China
became a net oil exporter by delivering 1 million tons of crude oil to
Japan. In 1974 China exported 4 million tons of crude oil to Japan
and another 1 million tons to Hong Kong, Thailand, and the Philippines. Thus, business firms in the United States and Japan view
China's potential of becoming a major exporter of crude oil and
petroleum products as providing the key for balancing her future
foreign trade.

CHINA AS AN EXPORTER OF CRUDE OIL

China's capability of exporting crude oil to other countries will
primarily be determined by the growth of output of crude oil in relation to the growth of domestic oil consumption. For China to continue to be a net oil exporter, the former must exceed the latter.

In terms of natural endowment China possesses a promising
energy potential, to the extent that resources should pose no serious
limitation on production levels in the near future. As Chapter 2
indicated, in 1952-74 China's crude oil output grew at an average
annual rate of 25.2 percent. In 1957-74 the annual rate was 24.5
percent, and in 1965-74 it was 21 percent. (See Table 2.9.) As the
production base enlarges, the growth rate generally tends to decline.
If past patterns are indicative, one might reasonably assume an

annual growth rate of 20 percent for crude oil output in 1975-77, 17
percent in 1978-80, 15 percent in 1981-83, and 12 percent in 1984-85.
Acceptance of these assumptions will yield a projection for China's
1980 crude oil output equal to 170 million tons and for her 1985 output
equal to 335 million tons. (See Table 2.13.)

The question that remains to be answered is, how will domestic
consumption respond to the growth of petroleum production? Will
China remain a coal-consumption economy, with 80 percent of her
fuel derived from coal, or will the Chinese economy gradually shift
to oil-consumption, as have those of Japan, Western Europe, and the
Soviet Union over the past 30 years?

Any attempt to forecast the consumption of petroleum in China
faces more difficulties than even the projection of output. Not only
are China's future consumption patterns hard to predict, but information on past petroleum consumption is unavailable. In early 1955 a
high ranking officer in the Chinese petroleum industry estimated that
by 1962, the end of the original Second Five-Year Plan, China probably would need to consume more than 10 million tons of petroleum
products.[4] This figure is based on the assumption that by 1962 China
would possess powered irrigation equipment rated at 15 million horsepower, which would consume 3 million tons of diesel oil. Furthermore, China in 1962 was assumed to possess 350,000 tractors for
agricultural use, at 15 horsepower per unit, which would consume
another 2 million tons of diesel oil. The balance of 5 million tons
would be needed in the form of gasoline, kerosene, and lubricants,
to be used for transportation, household lighting, and industrial uses.
Based upon this single estimation, and adopting the assumption that
consumption in one year is equal to domestic output plus the net
balance of imports and exports in a given year, K. C. Yeh has derived a series of petroleum consumption figures for the years 1955-62,
which are reproduced in Table 8.2.

Table 8.2 indicates that total consumption of major petroleum
products increased very rapidly during the 1955-62 period. Between
1952 and 1957, total petroleum consumption rose 2.2 times, from
.7 million tons to 2.27 millions.[5] From 1957 to 1962, total petroleum consumption rose threefold, the average annual rate of increase
being about 32 percent.

The major petroleum-consuming sectors in China today are
agriculture, industry, transportation, the household, and defense.
While consumption by the defense sector is extremely difficult to
estimate, demand for petroleum by the four civilian sectors between
1962-72 can be roughly estimated as follows:

Agriculture. In the agricultural sector, diesel oil is mainly consumed by tractors and irrigation equipment. In 1952 China possessed

TABLE 8.2

Estimated Petroleum Consumption, 1955–62

Product	1955	1956	1957	1958	1959	1960	1962
Total (in millions of tons)	1.82	2.10	2.27	3.08	4.22	4.95	9.00
Gasoline	0.94	1.06	1.01	1.28	2.17	2.36	3.00
Kerosene	0.39	0.46	0.60	0.68	0.87	1.09	1.50
Diesel oil	0.33	0.48	0.51	0.86	0.84	1.11	4.00
Lubricants	0.12	0.12	0.15	0.26	0.33	0.39	0.50
Total (in percent)	100.00	100.00	100.00	100.00	100.00	100.00	100.00
Gasoline	52.81	50.00	44.49	41.56	51.54	47.68	33.33
Kerosene	21.91	21.70	26.43	22.08	20.67	22.02	16.67
Diesel oil	18.54	22.64	22.47	27.92	19.95	22.42	44.44
Lubricants	6.74	5.66	6.61	8.44	7.84	7.88	5.56

Source: K. C. Yeh, Communist China's Petroleum Situation (Santa Monica, Calif.: Rand Corporation, 1962), p. 48.

a mere 2,000 standard units of tractors (15 horsepower for each
unit). The numbers increased to 24,629 units in 1957 and 154,200
units in 1965.[6] Output of standard tractors in China has risen rapidly
in recent years; the total number of tractors in operation by 1972 is
estimated at 354,000 units. (See Table 8.3.)

In addition, starting from 1964, China also produced garden
tractors in quantity. The garden tractor produced in the greatest
volume probably has been a model with a brake horsepower of seven.
Assuming a drawbar horsepower of four, one physical unit is equivalent to about one-fourth of a standard 15-horsepower unit. The rise
in output for the garden tractor has been phenomenal. Output in 1972
was officially reported as 24 times that of 1965, and 1973 output was
32 times that of 1965.[7] Total tractors in operation, including both
standard and garden tractors, when converted into 15-horsepower
units, approximates 405,000 units for 1972. (See Table 8.3.)

In the 1950s only a very small amount of irrigation equipment
was in use, but in 1957, the end of the First Five-Year Plan, China
possessed mechanized irrigation equipment utilizing 560,000 horsepower. The quantity of irrigation equipment, measured in terms of
horsepower utilized, reached 6.2 million in 1962 and 7.28 million in
1965.[8] During 1965-72, production of mechanized irrigation equipment rose more than three times, and by 1972 an estimated 24 million horsepower was in operation. (See Table 8.4.)

Early in 1958 an official document revealed the consumption of
diesel oil as 6 tons per year per tractor (15-horsepower unit) and
2 tons per year per horsepower of irrigation equipment.[9] Total
consumption for the 405,000 tractors and for the irrigation equipment
rated at 24 million horsepower therefore amounted to approximately
7.2 million tons in 1972.

Industry. In the past, petroleum consumption in the industrial sector
has expanded at approximately the same rate as industrial production.
As Chinese industry has become more mechanized, however, consumption of petroleum in the industrial sector should rise at a rate
higher than that of industrial production. China's industrial output
rose 132 percent between 1962 and 1972.[10] If we assume that the
growth rate of industrial consumption of petroleum was twice the
growth rate of industrial output, domestic industrial consumption of
petroleum in 1972 would have been 2.1 million tons.

Transportation. In the transportation sector, trucks and aviation use
gasoline and jet fuel, while steamships and some locomotives use
diesel oil.

TABLE 8.3

Tractor Production and Inventory, 1949-73
(in 15-horsepower units)

| Year | Standard | | Garden* | | Total |
	Production	Inventory	Production	Inventory	Inventory
1949	--	401	--	--	401
1950	--	1,286	--	--	1,286
1951	--	1,410	--	--	1,410
1952	--	2,006	--	--	2,006
1953	--	2,719	--	--	2,719
1954	--	5,061	--	--	5,061
1955	--	8,094	--	--	8,094
1956	--	19,367	--	--	19,367
1957	--	24,629	--	--	24,629
1958	957	45,330	--	--	45,330
1959	4,870	59,000	--	--	59,000
1960	22,000	79,000	--	--	79,000
1961	18,000	94,000	--	--	94,000
1962	8,000	100,000	--	--	100,000
1963	9,600	139,000	--	--	139,000
1964	11,800	149,000	150	150	149,150
1965	21,000	154,000	875	1,025	155,025
1966	32,000	150,000	2,625	3,650	153,650
1967	27,000	NA	2,100	5,750	NA
1968	30,000	NA	2,675	8,425	NA
1969	40,000	NA	3,200	11,625	NA
1970	70,000	272,000	9,000	20,625	292,625
1971	105,000	NA	9,625	30,250	NA
1972	115,000	354,000	21,000	51,250	405,250
1973	138,000	NA	28,000	79,250	NA

NA = Not available.

*The garden tractors are converted from physical units into 15-hoursepower units at a ratio of 4 to 1.

Sources: Figures for 1949-65 are from Chu yuan Cheng, The Machine-building Industry in Communist China (Chicago: Aldine-Atherton, 1971), pp. 204-5, Table 9-5; figures after 1965 are from Central Intelligence Agency, "Production of Machinery and Equipment in the People's Republic of China," May 1975, pp. 13-14.

TABLE 8.4

Production and Inventory of Powered
Irrigation Equipment, 1949-73

Year	Production	Inventory
1949	--	97
1951	--	118
1955	--	338
1956	170	508
1957	52	560
1958	720	1,280
1959	1,255	2,535
1960	1,610	4,145
1961	700	4,845
1962	955	5,800
1963	640	6,440
1964	860	7,300
1965	1,150	8,450
1966	1,530	9,980
1970	NA	16,911
1971	3,089	20,000
1972	4,016	24,016
1973	5,984	30,000

NA = Not available.

Source: Central Intelligence Agency, "Production of Machinery and Equipment in the People's Republic of China" (Central Intelligence Agency, May 1975), p. 11.

1. Trucks. Between 1956 and 1972, domestic production of trucks has been estimated at 554,000 units. During this period, China imported 130,000 trucks of diverse types.[11] By subtracting the number of trucks that had been scrapped, trucks in use in 1972 numbered approximately 550,000 units. Since one truck consumes an average of 18 tons of gasoline per year in China,* total gasoline consumption by trucks in 1972 would be around 10 million tons.

---

*According to one official statement, "crude oil output in 1952 was 436,000 tons which was not adequate for 20,000 trucks operated in a single year" (Feng Ta-lin, Ti-i-ko Wu-nien Chi-hua Chung-kung yeh ho yun shu-yeh ti Chi-pen Chien-she [Peking, 1956], pp. 12-20). This would indicate that annual consumption per truck was 22 tons of crude or about 18 tons of refined products.

2. Civil Aviation. In 1971 there were approximately 400 aircraft in China's civilian fleet, more than half of which were single-engine AN-2s. An AN-2 can accommodate 13 passengers or 1,500 kilograms of cargo. Other aircraft in the civilian inventory included several Viscount turboprop aircraft acquired from Great Britain in 1963 and 1964 and some Soviet IL-18s, IL-14s, IL-12s, AN-12s, AN-24s, and YAK-18 trainers. A 1974 study estimates the number and types of aircraft, exclusive of AN-2s, for civilian use in China, as including 183 aircraft, which can carry from 27 persons to 202 persons. (See Table 8.5.) Total civil route coverage, measured in terms of unduplicated route kilometers, has increased to 68,300 kilometers (42,440 miles) in 1973, an average expansion of 2,372 kilometers (1,474 miles) per year since 1950.[12] According to one Taiwan source, gasoline consumption by Chinese civil aviation was estimated to be 250,000 tons in 1969.[13] Petroleum consumption by civilian aircraft in 1972 may have approximated 400,000 tons.

TABLE 8.5

Estimated Number and Types of Aircraft
for Civilian Use, December 31, 1973

Type	Number	Country of Manufacture	Passenger Capacity	Range (in miles)
Antonov AN-24	25	USSR	44	1,445
Boeing 707	10	United States	202	3,925
Handley Page Herald	2	United Kingdom	56	700
Hawker Siddley Trident 1E, 2E, 3B	38	United Kingdom	180	1,100
Ilyushin IL-12	5	USSR	27	1,240
Ilyushin IL-14	54	USSR-China	28	920
Ilyushin IL-18	14	USSR	81	2,300
Ilyushin IL-62	5	USSR	186	4,160
Lisunov Li-2	28	USSR	25	1,500
T.S. 62	4	USSR	28	1,450
Tupolev TU-124	2	USSR	60	1,550
Vickers Viscount	7	United Kingdom	52	1,760

Source: David Chambers, "Civil Aviation in the PRC," Current Scene, August 1974, p. 12.

3. Railroad. As late as 1965, more than 95 percent of Chinese locomotives were steam locomotives using coal as fuel. Since 1966 China has produced diesel locomotives at an increasing rate. By 1971 diesel locomotives replaced steam locomotives as the main product. In addition, China imported 30 diesel locomotives from West Germany in 1970 and 50 additional large diesel locomotives from France in 1971. By 1972 diesel locomotives accounted for approximately 15 percent of the 6,000 locomotives operating in China.[14] According to one official source, a 4,000-horsepower diesel locomotive consumes an average of 9,125 tons of diesel oil a year.[15] It follows that a reasonable estimate of the consumption of diesel oil by the railroads for 1972 would approximate 9 million tons.

4. Civilian Ships. In 1950 the gross tonnage of modern vessels in China was only 60,000 tons. Between 1952 and 1966, China's shipyards constructed a total of 1.2 million dead-weight tons of ships for civilian use.[16] The gross tonnage of vessels constructed and placed in service during 1966-70 reportedly was nine times that of 1961-65, which had been 338,000 tons, or about 3 million tons.[17] Total civilian ships in service in 1972 are estimated at 3.5 million tons, an amount of tonnage that might have consumed a total of 2 million tons of diesel oil a year.

The Household. In 1958 Chinese official planners set the goal for household petroleum consumption at .005 tons per capita by 1972. Generously accepting the figures at face value and assuming a population of 800 million, this rate would require a total household consumption of 4 million tons. In the past five years China has developed numerous small hydraulic stations in the rural areas, which has resulted in a substantial decrease in the demand for kerosene for rural lighting. Total kerosene consumption in 1972 may have been approximately 1.5 million tons.

The total petroleum products estimated to have been consumed in the four civilian sectors in 1972 are summarized in Table 8.6 and compared with those estimated by K. C. Yeh for 1962 in Table 8.7.

Tables 8.5 and 8.6 cover only petroleum products for civilian use, excluding those consumed by the defense sector. Since there is no accepted pattern by which to estimate consumption by the defense sector, some arbitrary assumption has to be made. One Taiwan estimate put the military consumption of petroleum products at 620,000 tons for peacetime and 2,240,000 tons for combat situations.[18] Kusumi Tadao, a Japanese author and military commentator, estimated that China's peacetime military consumption of petroleum products should not exceed 2 million tons a year.[19] Based on these experts' assessments, 1.8 million tons of petroleum products are assumed for the military sector in 1972. Total 1972 consumption for both the civilian and the defense sectors approximated 34 million tons.

TABLE 8.6

Estimated Consumption of Petroleum Products
by the Four Civilian Sectors, 1972
(in thousands of tons)

Sector	Consumption	Percentage
Agriculture	7,200	23
Irrigation equipment	4,800	15
Tractors	2,400	8
Industry	2,100	7
Transportation	21,400	66
Trucks	10,000	31
Civilian aviation	400	1
Railroads	9,000	28
Civilian ships	2,000	6
Household	1,500	4
Total	32,200	100

Source: Figures are derived in the second section of Chapter 8
of the text.

TABLE 8.7

Pattern of Petroleum Products Consumption, 1962 and 1972

Consumption	Total	Gasoline	Diesel Oil	Kerosene	Lubricant
In millions of tons					
1962	9.0	3.0	4.0	1.5	0.5
1972	32.2	10.4	18.2	1.5	2.1
In percent					
1962	100.0	33.3	44.4	16.7	5.6
1972	100.0	32.3	56.5	4.7	6.5
1972 over 1962					
(1962 = 100)	357	347	455	100	425
Annual growth					
rate (in					
percent)	13.6	13.3	16.4	--	15.6

Sources: Figures for 1962 are from K. C. Yeh, Communist
China's Petroleum Situation (Santa Monica, Calif.: Rand Corporation,
1962), p. 48 (see Table 8.2); figures for 1972 are from second section of Chapter 8 of the text (see Table 8.6).

In the process of refining, petroleum fuel loss will consume some crude oil. In advanced countries the loss is less than 1 percent, as in the case of the United States, while in the Far East and Oceania the rate is about 4 percent.[20] In China the loss was as high as 15.7 percent in 1952 and 9.8 percent in 1957.[21] With progress in refining technology and improvement of refining facilities, the rate of loss should decline significantly. Assuming a 5 percent loss rate for 1972, the production of 34 million tons of petroleum products required a crude oil supply of 36 million tons.

Apart from its energy use, crude oil is needed as raw material for synthetic resins (including plastics), synthetic rubber, textile fiber, insecticides, and acaricides. The expansion of these industries in China has been very rapid in recent years. (See Section 4 of Chapter 7.) The consumption of crude oil as raw material for industrial and agricultural chemicals is believed to have been about 4 million tons in 1972.

The total 1972 demand for crude oil may thus add up to approximately 40 million tons. With a reported output of 45 million tons, there was a surplus of 5 million tons, some of which was added to Chinese inventories.

Assuming that the growth of consumption for 1975-85 will continue at the same rate as in 1962-72, that is, at an annual rate of 13.6 percent, we can roughly project China's petroleum consumption between 1974 and 1985 as shown in Table 8.8.

Table 8.8 shows the projected consumption and the projected export potential of Chinese crude oil during the 1975-85 period. Since the projection is based on past trends, any change in consumption patterns or production performance could have a significant impact on crude-oil export levels. As noted previously, one crucial factor that may affect the projection is whether the Chinese economy will continue to derive 70-80 percent of her energy from coal or instead gradually shift to relatively greater dependence on oil. There is evidence that the Chinese economy may move toward relatively greater oil consumption. In the 1952-73 period, when petroleum output increased 130 times, coal output rose only 5.6 times. The average annual growth rate of coal output was 8.7 percent, in contrast with 26 percent for oil. Moreover, expansion of coal output did not occur at a steady pace; the growth rate has fallen steadily, from nearly 15 percent during 1952-57 to 6.1 to 11 percent in 1957-65 and only 6 to 7 percent between 1965-73. In the richest and most accessible coal veins in the Manchuria area, coal reserves have been continuously depleted. To maintain an annual supply of coal from this region, it will be necessary to extract coal from poorer and less accessible seams in existing mines. Obviously these increasing costs will result in higher prices. Since the early 1950s the price of

TABLE 8.8

Projected Domestic Consumption and Trade Potential in Petroleum Products and Crude Oil, 1974-85
(in millions of metric tons)

Item	1974	1975	1976	1977	1978	1979	1980	1981	1982	1983	1984	1985
Total consumption of petroleum products	43.7	49.7	56.4	64.1	72.8	82.6	93.9	106.7	121.2	137.7	156.3	177.7
Petroleum products for civilian use[a]	41.6	47.3	53.7	61.0	69.3	78.7	89.4	101.6	115.4	131.1	148.9	169.2
Petroleum products for defense[b]	2.1	2.4	2.7	3.1	3.5	3.9	4.5	5.1	5.8	6.6	7.4	8.5
Refinery fuel and loss[c]	2.0	2.1	2.3	2.6	2.9	3.3	3.8	3.2	3.6	4.1	4.7	5.3
Total domestic demand for crude oil	50.3	57.0	64.6	74.7	84.8	96.2	109.7	126.4	143.5	163.1	185.2	210.5
Crude oil needed for fuel[d]	45.7	51.8	58.7	66.7	75.7	85.9	97.7	109.9	124.8	141.8	161.0	183.0
Crude oil as industrial raw material[e]	4.6	5.2	5.9	8.0	9.1	10.3	12.0	16.5	18.7	21.3	24.2	27.5
Estimated crude oil output	63.0	76.0	91.0	109.0	128.0	150.0	176.0	202.0	232.0	267.0	299.0	335.0
Surplus in crude oil[f]	12.7	19.0	26.4	34.3	43.2	53.8	56.3	75.6	88.5	103.9	113.8	124.5

[a] Assuming an annual growth rate of 13.6 percent, using 1972 consumption of 32.2 million tons as basis.

[b] Assuming 5 percent of civilian use.

[c] Assuming 4 percent rate between 1974 and 1980 and 3 percent between 1981 and 1985.

[d] Total consumption plus refinery fuel and loss.

[e] Assuming 10 percent of fuel consumption between 1972 and 1976, 12 percent between 1977 and 1980, and 15 percent between 1981 and 1985.

[f] Crude oil output less total domestic demand.

Sources: Figures are evolved as described in the second section of Chapter 8 of the text; output figures are from Table 2.13 (sources for output projections are in Table 2.13).

TABLE 8.9

Changing Structure of Primary Energy Consumption, 1951–73

Coal Equivalent

Year	(in millions of metric tons)					(in percent)			
	Coal	Crude Oil and Oil Products	Natural Gas	Hydroelectricity	Total	Coal	Crude Oil and Oil Products	Natural Gas	Hydroelectricity
1951	41.8	0.9	0.01	0.09	42.8	97	2	--	1
1953	65.8	2.1	0.01	0.3	68.2	96	3	--	1
1960	204.0	11.7	2.6	1.1	219.4	93	5	1	1
1961	124.0	10.9	3.8	1.0	139.7	89	8	3	--
1965	173.9	9.6	15.0	1.0	199.5	87	5	8	--
1970	235.2	24.7	23.2	1.5	284.6	83	9	8	--
1971	249.4	31.9	31.7	1.8	314.8	79	10	10	1
1973	269.5	61.6	59.9	2.6	393.0	69	16	15	--

Note: The conversion rate for raw coal mine production is .8; for raw coal local pit production, .5; for crude oil, 1.3; for refined products, 1.5; for natural gas, 1.332 per 1,000 cubic meters; and for electricity, .125 per 1,000 kilowatt-hours.

Source: Vaclav Smil, "Energy in the PRC," Current Scene 13, no. 2 (February 1975): 11.

coal has been rising steadily, from an estimated 12.5 yuan per ton in 1952,[22] to 25 yuan per ton in 1963.[23] During the same period the oil price has tended to be adjusted downward. Crude oil prices declined from 160 yuan a ton in 1952 to 58 yuan a ton in 1965.

According to a study made by an expert in Taiwan, the energy sources in China during the 1952-66 period showed a continued shift from coal to petroleum, as shown below in percentage:[24]

Year	Coal	Petroleum
1952	94.64	3.41
1960	89.61	7.49
1966	80.36	15.27

More recently, a study by Vaclav Smil also shows the same trend. (See Table 8.9.)

However, in view of the skyrocketing prices of crude oil in international markets, the shift from coal to relatively greater oil consumption in Chinese economy, if it does occur, is likely to be rather slow. It took two decades for crude oil and its products, which accounted for 3 percent of China's energy consumption in 1953, to increase its share to 16 percent in 1973. It may take until the 1990s for oil to account for one-third of Chinese primary energy consumption; consequently, the higher production costs of coal may not significantly alter the projections in Table 8.8 for 1974-85.

CHINA AS AN IMPORTER OF PETROLEUM EQUIPMENT

The petroleum industry is both resource-intensive and capital-intensive. The increase of crude oil output from an estimated 76 million tons in 1975 to a projected 335 million tons in 1985 is 4.2 times the increment achieved between 1953 and 1974, when crude oil output was raised from .6 million tons to 63 million tons. During the 1953-74 period, fixed capital investment in the petroleum industry was estimated as 21 billion yuan, of which 10.6 billion, or approximately 50 percent, was allocated for petroleum equipment and machinery, while 2.4 billion yuan (or 1.2 billion dollars) was imported from abroad. (See Table 5.1.) If the past relationship between capital investment and output is assumed to remain unchanged, China will need to invest 88 billion yuan in the petroleum industry as fixed capital in order to produce the projected 335 million tons by 1985. Of this 88-billion-yuan investment, 44 billion yuan will be allocated for equipment and machinery, including 8.8 billion yuan for equipment that will be imported from abroad. At the 1974 official exchange rate of 1 dollar to 1.9739 yuan, this would amount to an aggregate acquisition of $4.5 billion. (See Table 8.10.)

TABLE 8.10

Projected Fixed Capital Investment in Petroleum Industry
and Demand for Petroleum Equipment, 1975-85
(in millions of yuan*)

Period	Fixed Capital Investment in Petroleum Industry	Petroleum Equipment			
		Demand	Domestic Supply	Import	Self-Sufficiency Rate (in percent)
1953-74	21,051	10,562	8,123	2,439	76
1975-85	88,400	44,200	35,400	8,800	80

*One yuan = $.507.

Sources: Figures for 1953-74 are from Table 5.1; figures for
1975-85 are from text.

The projected need to import an increasing amount of foreign
equipment is a consequence of the fact that a large portion of the
projected crude oil output for the 1975-85 period will come from offshore resources. Recent official reports confirmed that both the
Shengli oil field and the Takang oil field are now engaged in offshore
drilling.[25] Although onshore drilling and production of crude oil are
well within the limits of current Chinese technology, the extraction
of offshore petroleum still possesses serious technical problems
that will require the help of foreign equipment and know-how.

Technological barriers to the growth of crude output may also
arise in the areas of refining and transportation. As noted in the
second section of Chapter 4, there has been a widening gap between
the increase in crude oil production and the growth of refining capacity. Given the present capacity and constraints of the Chinese petroleum equipment industry, rapid expansion of refining capacity may
not be feasible without a major increase in the supply of foreign refining equipment.

Equally important to the growth of crude output is the expansion
of transport facilities. If China is to produce and export large quantities of crude oil, a massive program of pipeline construction and
expansion of both tanker fleets and harbor facilities is a necessity.

These factors, along with the best available evidence, indicate
that an increasing importation of petroleum equipment and transportation equipment is an indispensable condition if China is to become a major exporter of crude oil.

A relevant and poignant query at this juncture is what share of China's future importation of petroleum equipment is likely to be purchased in the United States. This can hardly be answered without first attempting to project Sino-American trade for 1975-80.

A 1974 study by Alexander Eckstein and Bruce Reynolds estimated that between 1972 and 1980 the average annual growth rate of China's foreign trade would be around 10 percent. This would mean that the total turnover in 1980 would be more than twice as large as in 1972. Moreover, this study contended that in contrast to past patterns, China's imports would tend to grow at a faster rate than its exports. Eckstein and Reynolds assume that the Chinese will have some medium-term credits at their disposal that will enable them to maintain a trade deficit. Given these assumptions, Eckstein and Reynolds projected that China's overall imports may rise to $7.6 billion in 1980 (in 1973 prices), with exports increasing somewhat more slowly.

Accepting the above general framework and further assuming that the United States will supply about 20 percent of China's imports, it follows that U.S. exports to China would be about $1.5 billion in 1980, approximately twice the 1973 level.[26]

By extending Eckstein's projection into 1985 and assuming an average annual growth rate of 10 percent, China's overall imports in 1935 can be extrapolated to $12.2 billion (at 1973 prices). Assuming again that the United States maintains a 20 percent share of China's total imports, it follows that U.S. exports to China in 1985 should range near $2.4 billion.

Of the U.S. produce and commodity exports to China, Eckstein and Reynolds foresee a decline in farm products and a protracted rise in machinery and equipment exports.[27] Accepting their assumptions, but extending their forecast period from 1980 to 1985, the product and commodity composition of U.S. exports to and imports from China can be summarized in Table 8.11.

Based on the discernible trends in the composition of trade shown in Table 8.11, one can reasonably estimate that the United States may be exporting to China machinery and equipment valued at $840 million (at 1973 prices) by 1985. Of this total, from 30 to 50 percent will probably be in the field of petroleum equipment. In monetary terms, the United States may export $150 to $225 million of petroleum equipment in 1980 and $280 to $420 million in 1985. Aggregate exports of U.S. petroleum equipment during 1975-85 will probably be of a magnitude of $1.5 to $2 billion.

With respect to these prospective petroleum equipment imports, the Chinese will most likely give priority to the following items:

TABLE 8.11

Commodity Composition of U.S. Exports to and Imports from China
in 1973 and Projections for 1980 and 1985
(in percent)

Commodity Category	Export			Import		
	1973	1980	1985	1973	1980	1985
Food and live animals	62.2	40	35	9.3	15	15
Beverages and tobacco	0.2	--	--	0.6	--	--
Crude materials, inedible, except fuels	25.5	10–15	10	25.1	26	30
Mineral fuels, lubricants, and related materials	--	--	--	0.6	--	--
Animal and vegetable oils and fats	4.1	--	--	1.3	3	3
Chemicals	0.6	5–10	5	12.5	--	--
Manufactured goods	1.1	5–10	10	30.1	31	32
Machinery and equipment	6.0	30	35	0.6	--	--
Miscellaneous manufactured articles	0.1	5	5	18.1	15	15
Other	--	5	5	1.3	10	5

Sources: Figures for 1973 and 1980 are from Alexander Eckstein and Bruce Reynolds, "Sino–American Trade Prospects and Policy," American Economic Review, May 1974; 1985 is estimated by the author as described in Chapter 8.

1. Offshore drilling equipment, including rigs, tugs, and platforms. China has purchased only three offshore drilling rigs, secondhand, from Japan. If large-scale exploitation of oil resources on the Pohai Gulf is to be undertaken, China may soon have to look toward the United States for more powerful and advanced equipment. American companies have dominated the offshore drilling industry in the past and are likely to continue doing so in the foreseeable future. According to Oil and Gas Journal in 1973, of the 200 mobile rigs operating in the world at the end of 1972, over 160 were owned by American companies; of the remaining 40, ten were co-owned and operated by U.S. firms.[28] Thus about 85 percent of the world's mobile fleet of oil rigs were completely or partially American-owned. The same source also predicted that by 1975 U.S. firms will effectively control at least 80 percent of all offshore drilling rigs. Preeminently established as the world's leading supplier of offshore drilling equipment, the United States is in an excellent position to garner a large share of China's purchases of offshore rigs and associated equipment.

2. Deep-well drilling rigs. China can now supply onshore drilling rigs capable of penetrating 3,200 meters. For deep-well drilling beyond this limit, the equipment has to be imported.

3. Steel pipes for pipeline. Steel pipes produced in China are far short of domestic demand. In the early 1970s China has continued to import steel pipes from Japan and from West Germany. The largest pipe China currently produces has a diameter of 30 inches (75 centimeters). For pipes with larger diameters China still relies on imports.

4. Compressors and pumps. China can now produce cylinders capable of withstanding 1,500 to 2,000 atmospheres of pressure. However, China still finds it necessary to import high-pressure compressors and pumps capable of handling 5,000 to 7,000 atmospheres of pressure.

5. Sophisticated equipment for exploratory drilling. Although China is now capable of producing a wide range of exploration machines, her petroleum industry still needs more sophisticated equipment. The current purchase of $5 million in seismic equipment from Geo. Space Corporation of Houston, Texas, exemplifies the need in this field.

6. Blowout-preventer equipment. Chinese official reports reveal a high blowout accident rate. Chinese domestically made equipment has apparently failed to control pressure in exploratory oil and gas wells effectively. It is probable that China will seek to purchase the more advanced Western blowout-preventive equipment.

7. Petrochemical plants. Since the ultimate goal of the Chinese petroleum industry is not to export hugh quantities of crude but

to expand China's petrochemical industries, the procurement of
petrochemical plants is likely to continue for several years. Plants
associated with ammonia-fertilizer manufacturing are particularly
in demand.

In general, the Chinese market for U.S. oil-exploration equipment and petrochemical plants appears to be very promising at least
through 1985.

SINO-AMERICAN TRADE IN PERSPECTIVE

However, the prospects for Sino-American trade cannot be
assessed solely from an economic perspective, any more than they
can be viewed as isolated instances of bilateral trade. It is more
meaningful to view China's trade relations as only a link in the entire
chain of international relations.

In a country like China, where politics takes command, economic decisions are seldom made without taking account of the political consequences. A survey of the economic history of the People's
Republic during the past 25 years clearly indicates that economic
and political considerations are closely intertwined. The current
efforts of China to expand her petroleum industry rapidly are moti-
vated by both economic and political considerations.

From the economic point of view, the Chinese authorities apparently view the tremendous profit potential of oil exports as a
golden chance to bolster the Chinese industrialization program. The
export of 4 million tons of crude oil to Japan in 1974 enabled China
to purchase modern integrate iron and steel mills and petrochemical
plants that would have cost more than 10 million tons of crude oil
a few years before. Perhaps more significant to the Chinese than
the economic motive are the strategic implications of the changing
balance of power in Asia and the Far East.

Ever since the Sino-Soviet border clash in 1963, China has
considered the Soviet Union her most immediate military threat. As
the Soviet military buildup escalated along the Sino-Soviet border,
the Chinese leadership decided that a detente with the United States
might provide the most credible deterrent to a Soviet preemptive
attack.

The fundamental change in Sino-American relations also affected Sino-Japanese and Soviet-Japanese relations. Aware of
Japan's economic and military potential, both China and the Soviet
Union energetically sought the cooperation of Japan. In this delicate
power alignment, the supply of crude oil to Japan became a key factor in the negotiations.

At the fifth Japanese-Soviet joint economic committee meeting in Tokyo in 1972, the Soviet government asked for Japanese bank loans totaling $1.51 billion to help finance a new $3 billion trans-Siberian pipeline program, as well as joint participation in developing Soviet coal and natural gas resources. In outlining the Soviet terms for the joint exploration venture of the Tyumen oil field, the Soviet government assured the Japanese that the USSR would be able to supply Japan with 25 million to 40 million tons of low-sulphur crude oil per year for the next 20 years.[29] The Soviet proposal, if consummated, would draw Japan closer toward Moscow. Moreover, the pipeline would skirt the periphery of China's northern border. From China's prospective, the pipeline would increase the strategic mobility of the Soviet Union to war against China.

It was probably to forestall the impending Soviet-Japanese cooperation that China normalized her diplomatic relations with Japan in September 1972 and has subsequently ventured petroleum offerings to Japan. In 1973 China delivered 1 million tons of low-sulphur crude oil of the highest quality to oil-poor Japan. In 1974 China supplied Japan with 4 million tons of crude oil. Recent reports indicate that China's crude exports to Japan may reach 10 million tons in 1975. If this trend continues, China may supply Japan with up to 40 million tons by 1979, which would tend to make the Soviet offer less attractive.

Strategic considerations would imply that if China is to export crude oil in large quantities, the chief recipient would tend to be Japan rather than the United States. There are also some cogent nonpolitical factors favoring this opinion.

First, geographic proximity results in relative low transportation costs between China and Japan. For Japan, Chinese crude oil supplied in large quantities would constitute the most logical substitute for crude from the Middle East.

Second, the crude oil produced at Tach'ing has a low sulphur content, which is particularly desirable for the densely populated and highly industrialized Japanese environment.

Third, the crude oil produced at Tach'ing and Takang has a very high wax content that makes processing extremely difficult in refineries not specifically equipped to handle it. The shipment of Chinese crude to Thailand was suspended in the autumn of 1975 due to the fact that refineries in that country are unable to handle it. According to experts' opinion, the Takang crude had a pour point of 90 degrees compared with a pour point of 10 to 20 degrees for most Middle Eastern crude. To handle this oil it must constantly be kept heated and its best temperature is 130 degrees. Yet, the Takang crude also has a relatively low flashpoint approaching the 130-degree temperature so that heating it to the temperature at which it will flow easily would raise the risk of fire or explosion.[30] In the Far East area probably only Japan has the technology and facilities to handle the Chinese crude in large scale.

Despite these factors supporting the hypothesis that China will export most of her oil surplus to Japan, there are reasons for believing that China may export a portion of this crude to other countries, including the United States. One of the fundamental tenets guiding China's foreign trade since the early 1960s has been to avoid overdependence on any particular country. In buying machinery and equipment in 1965-74, China placed orders with more than ten countries, thereby preventing any one of them from dominating her trade. The same principle may also apply to her oil exports in the future.

If China can export 50 million tons of crude oil in 1980 and 100 million tons in 1985, at the present price of $100 per ton her oil revenue would amount to $5 billion in 1980 and $10 billion in 1985. This could finance one-half of China's imports in both years, enabling China to expand her industrial base by purchasing more modern plants and equipment from Japan, the United States, and Western Europe. The allocation of Chinese imports among the different industrial powers would also be determined by political, economic, and technological factors. There are, however, several factors favoring import procurement from the United States.

First, U.S. technology in offshore operation, deep-well drilling, and refining is still unrivaled. The Chinese authorities in charge of the petroleum industry have a great respect for U.S. petroleum technology.[31]

Second, U.S. products dominated China's market for quite a long period in history. Despite the 21-year disruption, the Chinese have had a long working experience with American machinery and equipment. The working life and the strength of these types of American products are generally superior to the Japanese models, and this fact is not likely to go unnoticed by the Chinese authorities.

Third, as long as Sino-Soviet conflict persists, Sino-American economic cooperation is likely to be encouraged.

On the other hand, there are also factors favoring an expansion of Sino-Japanese trade. Apart from geographical proximity, Japan provides a wide range of capital goods that are less expensive, simpler to construct and maintain, and more suitable to the physical characteristics of the workers using them. Moreover, Japan in recent decades has become an advanced producer in the fields of shipbuilding, steel rolling, and electronics. It might be more advantageous for China to order offshore platforms, steel pipes, dredgers, and tankers from Japan than from the United States. Furthermore, it was the United States that embargoed trade with China for 21 years, while the Japanese never suspended their trade relations with the

People's Republic. The Japanese enterprises and government are more familiar with and experienced in handling Chinese contracts. Assuming these favorable factors continue in operation, the Japanese forecast increases in Sino-Japanese trade to $3,450 million in 1975, $5,050 million in 1977, and $10 billion in 1982.[32] Since Sino-Japanese trade already exceeded $2 billion in 1973 and approached $3 billion in 1974, the forecast for the 1980s may well be realized. This provides some evidence that during the next decade Japan will remain unchallenged as China's number-one trading partner.

In conclusion, and attempting to take all these related factors into account, it is likely that the United States will be second only to Japan as a trade partner with China. Between 1975 and 1985 Japan may account for 30 percent of China's total trade, while the United States, Western Europe, and the Third World will each account for approximately 20 percent, with the remaining 10 percent going to the communist bloc. For the United States, total trade with China will probably rise from $934 million in 1974 to 4.8 billion in 1985, a fourfold increase in 11 years. Exports from the United States to China may reach $2.4 billion in 1985, of which petroleum equipment and machinery may account for from $280 to $420 million.

By 1985 China may export 100 million tons of crude oil to Japan and other countries. The pattern of Chinese trade will become more multilateral in nature. China's gaping deficits in trade with the United States will then be covered by oil revenue from Japan, Southeast Asia, and elsewhere.

To produce 100 million tons of crude surplus in 1985, the Chinese petroleum industry will need to import a substantial amount of foreign equipment and technology. Past experience suggests there will be little prospect for foreign oil companies of entering into joint ventures or even of operating under service contracts. Prospects are even less that the People's Republic will sacrifice the principle of no foreign participation for the expeditious development of her petroleum industry.

NOTES

1. Chu-yuan Cheng, The Chinese Market under Communist Control (Hong Kong: Union Research Institute, 1955), p. 3.

2. Alexander Eckstein, "China's Economic Growth and Foreign Trade," U.S.-China Business Review, Washington, D.C., July-August, 1974, pp. 15-18.

3. New York Times, June 3, 1975.

4. Liu Fan, assistant to the Minister of Petroleum Industry, "Report to the First National Conference on Local Petroleum Industry," Shih-yu Lien-chih, no. 6 (1958), p. 1.

5. K. C. Yeh, Communist China's Petroleum Situation (Santa Monica, Calif.: Rand Corporation, 1962), p. 48.

6. Chu-yuan Cheng, The Machine-building Industry in Communist China (Chicago: Aldine-Atherton, 1971), p. 259.

7. Shih Cheng, A Glance at China's Economy (Peking: Foreign Languages Press, 1974), p. 23; also China Reconstructs, January 1975, p. 6.

8. Cheng, op. cit.

9. Shih-yu Lien-chih, no. 6 (1958), p. 1.

10. This figure is based on U.S. Congress, Joint Economic Committee, People's Republic of China, An Economic Assessment, 92nd cong., 2nd sess. (Washington, D.C.: Government Printing Office, 1972), p. 63.

11. Ibid., pp. 163-64.

12. David Chambers, "Civil Aviation in the PRC," Current Scene, August 1974, pp. 4-5.

13. Hsiao Chi-jung, Chung-kuo-ta-lu chih Neng-yuan yu Li-yung (Taipei: Ministry of Economic Affairs, 1970), p. 236.

14. Central Intelligence Agency, Production of Machinery and Equipment in the People's Republic of China (Washington, D.C.: Central Intelligence Agency, May 1975), p. 16.

15. Hsiao Chi-jung, op. cit., p. 235.

16. Cheng, The Machine-building Industry in Communist China, op. cit., p. 257.

17. U.S. Congress, op. cit., pp. 168-69.

18. Shih Chien, "A Study of the Nature and Amount of the Defense Expenditure in Communist China," Chung-kuo Ta-lu Yen-chiu [Mainland China Studies] (Taipei), no. 5 (May 10, 1971), p. 20.

19. Kusumi Tadao, "Communist China's Military Capability," translated by Institute of Political Research, Taipei, in Ti-ch'ing yen-chiu [Studies of Enemy Affairs], no. 398 (August 1968), p. 9.

20. This figure is derived from the statistics in George Sell, The Petroleum Industry (London: Oxford University Press, 1963), Appendix 3, p. 272.

21. The First Five-Year Plan for Development of the National Economy of the People's Republic of China, 1953-1957 (Peking: Foreign Languages Press, 1956), p. 28.

22. Ta-chung Liu and Yeh Kung-chia, The Economy of the Chinese Mainland; National Income and Economic Development (Princeton: Princeton University Press, 1968), p. 569.

23. Thomas G. Rawski, "The Role of China in the World Energy Situation," unpublished paper (Washington, D.C.: The Brookings Institution, 1973), p. 15.

24. Hung Yu-ch'iao, "An Appraisal of the Changing Fuel Structure on the China Mainland," Issues and Studies (Taipei), (Taipei), November 1974, p. 66.

25. Peking Review, no. 41 (October 11, 1974), pp. 18-19.

26. Alexander Eckstein and Bruce Reynolds, "Sino-American Trade Prospects and U.S. Policy," The American Economic Review, May 1974, pp. 294-99.

27. Ibid., p. 3.

28. Oil and Gas Journal, December 31, 1973, pp. 136-37.

29. New York Times, February 25, 1972, pp. 53-54.

30. New York Times, September 19, 1975.

31. The Wall Street Journal, September 6, 1974, p. 20.

32. China Trade Report, August 1972, p. 3.

SUMMARY OF THE STUDY

The findings of this study can be summarized as follows:

1. The record of the Chinese petroleum industry in the past 25 years, when contrasted with the records of other modern industries of the People's Republic of China, has been phenomenal. Between 1952 and 1974, the average annual growth rate of crude oil output reached 25.4 percent, as compared with 14.4 percent for crude steel output and 8.4 percent for the overall industrial output, during the same period. As a result, the relative share of the petroleum industry in eight basic industries advanced from 1 percent in 1957 to 3.6 percent in 1965, 7.4 percent in 1972, and 8.2 percent in 1974, a position surpassed only by the machine-building and chemical industries.

In terms of tonnage, China's crude oil output increased from less than .5 million tons in 1952 to 5.2 million tons in 1960. The 1960 output was doubled by 1965, which in turn redoubled to 20.7 million tons by 1969. Even more impressive is the tripling of crude oil output between 1969 and 1974. By 1974 output was equivalent to 63 million tons, an amount approximately equal to Soviet output in 1954 (see Table 9.1).

2. The most significant development since the early 1950s has been the opening of three major oil fields in the coastal area: Tach'ing in North Manchuria; Shengli in the lower Yellow River region extending to Pohai Gulf; and Takang in the Peking-Tientsin area. In 1974 Tach'ing produced 20.74 million tons of crude oil, or 37 percent of the nation's total; Shengli produced 10.21 million tons, or 19 percent of the nation's total; and Takang produced 4.12 million tons, or 7 percent of the nation's total. Thus, in aggregate, these three oil fields turned out 63 percent of China's crude oil in 1974.

TABLE 9.1

Major Indicators of the Chinese Petroleum Industry, 1952-74

Indicator	1952	1957	1960	1962	1965	1970	1972	1974
Crude output (in millions of tons)	0.44	1.46	5.20	6.70	11.00	29.10	45.00	63.00
Probable oil reserves (in millions of tons)	NA	200	400	NA	1,000	2,000	NA	3,000
Gross petroleum output value (in millions of 1952 yuan)	240	810	2,910	3,740	6,140	16,260	25,080	35,770
Petroleum output value as percent of total	0.6	1.0	1.5	2.0	3.6	6.0	6.4	8.2
Employment in petroleum industry (per thousand workers)	22	67	NA	NA	NA	NA	NA	1,162
Imports of crude and petroleum products (in millions of tons)	0.61	1.80	3.27	1.95	0.28	--	--	--
Exports of crude and petroleum products (in millions of tons)	--	--	--	--	--	--	1.0	5.0
Civilian consumption of petroleum products (in millions of tons)	NA	2.27	4.95	9.00	NA	NA	32.2	41.6

NA = Not available.

Sources: Figures are derived from various sources as shown in Tables 2.5, 7.1, 7.2, 7.4, 7.8, and 8.8.

These three new oil fields have not only accelerated the forward thrust of the petroleum industry, thereby enabling China to become a net exporter of oil, but have also fundamentally transformed
China's oil distribution. Previously there was a critical imbalance
in the regional distribution of the Chinese petroleum industry:
whereas 80 percent of the country's industrial output was produced
in the coastal areas, more than 90 percent of the oil resources were
concentrated in the remote areas of the northwest. By developing
these new inland and offshore oil resources located near the industrialized coastal areas, this imbalance was largely corrected. This
relocation of the industry effected tremendous savings in transportation costs and also afforded the opportunity for large-scale exports.

   3. Data on China's oil reserves are inadequate, to say the
least. Although continuous prospecting during 1965-74 led to the discovery of several promising oil fields, reasonably trustworthy data
on these discoveries are unavailable.

   Based on scattered and fragmentary evidence, most China experts think China's total oil reserves, including the recent discoveries of offshore deposits, are over 30 billion tons, or 220 billion
barrels. Of this total, only 3 billion tons, or 22 billion barrels, can
be counted as probable reserves.

   Assuming that crude oil will maintain a 20 percent annual
growth rate between 1975 and 1977; a 17 percent rate between 1978
and 1980; a 15 percent rate between 1981 and 1983; and a 12 percent
rate between 1984 and 1985, the current estimated probable oil reserves could sustain these rates of exploitation for at least 30 years.
Additionally, the continuous exploration of new fields should help to
insure China's ability to maintain adequate oil reserves into the
twenty-first century.

   4. The refining capacity of China exceeded crude oil production in the early years of the petroleum industry development,
1952-62, and managed to keep pace with crude oil growth until 1970.
Subsequently, a gap between these two measures developed whereby
in 1974 Chinese refineries could process only 75 percent of the crude
oil produced. It has been projected that by 1985, even assuming a
significant expansion of refining, China will be able to process only
70 percent of her crude oil.

   Transportation facilities will be another future bottleneck for
the Chinese petroleum industry. In most major oil-producing countries, crude supplies are moved to refineries by a system that com-
bines pipelines and ocean tankers. Because one-third of the crude
oil produced in China is inaccessible to waterways and because of
the comparatively high costs associated with the construction and
maintenance of a pipeline system, the transportation of oil still relies largely on railroads and highways. The costs of such primitive

transport are prohibitive for some areas and have retarded the exploitation of oil fields in the hinterlands.

Although China's first pipeline was laid in 1958, only 5,000 kilometers of pipeline were in operation by 1975. In 1965-74 China expanded her tanker fleet, which had an estimated total of 400,000 tons by 1974. To facilitate future oil exports, major harbors along the eastern coast have been expanded and mechanized. Noteworthy have been the expansions of Dairen, Chinwangtao, Tientsin, Tsingtao, Shanghai, and Chankiang; these seaports will become the main depots for crude oil shipments to Japan, Southeast Asia, and elsewhere.

5. In the 21 years between 1953 and 1974, total capital investment in the Chinese petroleum industry was estimated at 21 billion yuan or approximately $9 billion, of which $4.5 billion was allocated for petroleum equipment and machinery. Machinery imported from abroad was estimated at about $1 billion. During this period 75 percent of the petroleum machinery needed was supplied by domestic producers.

Prior to 1954 there had been virtually no plants specializing in producing petroleum equipment, since most of the equipment was supplied by the Soviet Union. China began producing drilling machines in 1954 and greatly expanded her capacity in this field during the 1958-64 period. Specialized plants, as well as multiple-product factories, were established in Shanghai, Harbin, and Lanchow. By 1971 the petroleum equipment industry had become a major division among the 13 divisions in China's rapidly growing machine-building industry.

With the exception of offshore drilling equipment, petrochemical equipment, and some sophisticated instruments, China's petroleum equipment industry is now capable of producing a range of products wide enough to satisfy approximately 80 percent of the domestic demand.

6. Despite this impressive progress, in technology and variety the petroleum equipment produced in China still lags significantly behind what is produced by the advanced industrial nations. It may take China until the mid-1990s to attain the current production and technology levels of the West and Japan. During this interval China will continue to have to rely on foreign supplies, not only for integrated plants and equipment for offshore exploration, but also for equipment for deep-well drilling, petrochemical plants, large-diameter oil pipes, and large-size tankers.

During the early years of her petroleum industry development, China obtained most of the equipment and machinery from the Soviet Union. Between 1955 and 1960 China imported $123.8 million in oil-well drilling equipment and pipes from the Soviet Union. The Sino-Soviet split in 1960 resulted in the almost complete cessation of

Soviet supplies; by 1962 China's imports of machinery and equipment from the USSR were reduced to only 5 percent of the value of the goods imported in 1959.

7. To fill the hiatus caused by the virtual suspension of Soviet supplies, China turned to Western Europe and Japan for petroleum equipment and supplies. Initially very limited, these purchases from the West and Japan escalated rapidly after 1970, as the quest for petroleum extended from onshore to offshore exploration. Equally important in China's overseas procurement was the acquisition of petrochemical plants. In 1973-74 the aggregate value of plants contracted for in the petrochemical field amounted to $900 million, one-third of which were imported from Japan.

Corresponding to this extensive procurement of foreign equipment, China has also stepped up her acquisition of the latest technology from the West and Japan, as evidenced by the increased traffic of groups of technical experts between China and other lands. Likewise, new contracts contain provisions for technical data and the training of Chinese technicians.

In the process of acquiring foreign equipment, China began to accept medium-term commercial credit to finance her imports. Using deferred payments as a means of financing imports from non-Communist countries will increase China's ability to import and may establish a new pattern for trade relations with the free world.

8. Accompanying China's current rapid growth of oil production has been the resumption of trade relations with the United States. After 21 years of disruption, Sino-American trade was again initiated in the second half of 1971. In the following year the amount of U.S. exports to China totaled $60 million, while U.S. imports were $32 million. A vigorous advance took place in 1973, when the United States supplied China with $740 million worth of farm products, transportation equipment, and machinery, in exchange for $60 million in commodities from China, a twelvefold increase over the preceding year. This metamorphosis in trade relations made the United States second only to Japan as a trading partner of China.

The resumption of Sino-American trade relations has been characterized by a growing differential between U.S. exports to and imports from China. The ratio was 2:1 in 1972, 12:1 in 1973, and about 7:1 in 1974. Although China has financed her trade deficit with the United States through her surplus earned in Hong Kong and Southeast Asia, her ability to continue large-scale imports is affected by her export capability. Given current world oil prices, China can perhaps best promote her trade with the world by exporting crude oil. Thus the potential to export oil will enable China to balance her foreign trade in the late 1970s and early 1980s.

9. China's capability to export crude oil will mainly be determined by the growth of domestic output in relation to the growth of domestic oil consumption. Based on the assumption that China's crude oil output will grow at a diminishing annual rate (20 percent between 1975 and 1977, 17 percent between 1978 and 1980, 15 percent between 1981 and 1983, and 12 percent between 1984 and 1985), China will produce 176 million tons of crude oil in 1980 and 335 million tons in 1985.

Domestic consumption, on the other hand, is estimated to have been 50 million tons in 1974 and projected to be 110 million tons by 1980 and 210 million tons by 1985. This would result in a probable crude oil surplus of 56 million tons in 1980 and 125 million tons in 1985, a large portion of which may be exported.

10. The increase of crude oil output from an estimated 76 million tons in 1975 to a projected 335 million tons in 1985 will require an additional capital investment of $45 billion, half of which will probably be used for purchasing machinery and equipment. Of the $22.5 billion invested in new plant and equipment, perhaps $4.5 billion will be imported from abroad. Priority will probably be given to the following imports: offshore drilling equipment, deep-well drilling rigs, steel pipes for pipelines, compressors and pumps, sophisticated equipment for exploration, blowout-preventor equipment, and petrochemical plants.

11. Although China may export 50 million tons of crude oil in 1980 and as much as 100 million tons in 1985, most of the crude oil will probably be exported to Japan rather than to the United States. The reason for this policy is that such oil exports to Japan would serve to forestall the possible Soviet-Japanese joint exploration of the Tyumen oil field. The geographic proximity of China and Japan is a cost factor that also augurs favorably for such a policy. Moreover, China's adherence to the principle of independence of action virtually excludes foreign oil companies from entering into joint ventures with China to explore Chinese offshore regions.

Additionally, it would appear that the ultimate goal of China's petroleum development policy is not to supply foreign countries with huge quantities of crude but to expand her own petrochemical industry. This policy is evidenced by her extensive acquisition of petrochemical plants during recent years.

For all these reasons, the relative share of future Chinese oil exports going directly to the United States would appear to be rather small.

12. Chinese purchases of foreign-made petroleum equipment in recent years indicate that China clearly considers the exploitation of the oil resources under the continental shelf as a primary task to

serve the rapid development of the Chinese petroleum industry. The field of sophisticated offshore drilling rigs and equipment is one in which American products and technology command relative superiority, enabling the United States to compete with Western Europe and Japan for the Chinese market.

In the late 1970s and early 1980s the pattern of Chinese trade will become more multilateral. The gaping deficits in trade with the United States are likely to be covered by China's oil revenue from trade with Japan and Southeast Asia.

Specifically, this analysis leads to the conclusion that the Chinese market for U.S. oil exploration equipment and petrochemical plants appears to be quite favorable, at least through 1985.

INTERNATIONAL SIGNIFICANCE

The international role of China's petroleum industry can also be assessed by comparing her reserves, output, transport facilities, and pattern of energy consumption with those of other major oil-producing countries.

Although China's petroleum industry has recorded an impressive growth rate since the early 1960s, her relative share of the total world output remains exceedingly small. As shown in Table 9.2, China's crude output of 63 million tons accounted for only 2.3 percent of the world's total as of 1974, or 14 to 15 percent of the individual output of each of the world's top three oil producers, the Soviet Union (450 million tons), the United States (440 million tons), and Saudi Arabia (425 million tons). Among the world's 24 major oil-producing countries, China ranked thirteenth, a little above Algeria but lower than Indonesia.

However, the international significance of the Chinese petroleum industry lies in its long-term prospects, which are heavily dependent upon two major factors: her potential reserves and her energy structure. Although the probable reserves of Chinese petroleum constituted only four percent of the world's total proven reserves as of 1974, most of the Chinese oil-bearing basins remain basically untapped. In short, additional exploration, using sophisticated instrumentation and techniques, should substantially increase China's probable reserves. With respect to energy structure, unlike most industrial powers and many of the less-developed nations, which are highly dependent upon oil, China relies basically upon coal as an energy source. Table 9.3 indicates that in 1973 oil accounted for 74 percent of the energy sources in Latin America, 57.7 percent in the less-developed countries in the Eastern Hemisphere, 63.3 percent in Western Europe, 51.8 percent in Canada, 46.1 percent

TABLE 9.2

World Proven Reserves and Crude Output, 1974

Country	Proven Reserves (in billions of barrels)	(in percent)	(by rank)	Crude Output (in millions of tons)	(in percent)	(by rank)
Middle East						
Saudi Arabia	132.0	22.7	1	425	15.6	3
Kuwait	64.0	11.0	3	110	4.2	7
Iran	60.0	10.3	4	305	11.2	4
Iraq	31.5	5.4	6	100	3.7	8
United Arab Emirates	24.0	4.1	8	90	3.3	11
Qatar	6.5	1.2	16	25	0.9	16
Bahrain	0.4	0.1	25	3.5	0.1	23
Oman	5.3	0.9	20	15	0.5	19
North America						
Canada	9.0	1.5	14	100	3.7	10
United States	35.3	6.0	5	440	16.1	2
Mexico	5.4	0.9	19	35	1.3	15
Latin America						
Venezuela	14.0	2.4	12	145	5.3	5
Ecuador	5.7	1.0	18	10	0.3	21
Argentina	2.4	0.4	21	20	0.7	19
Asia						
China	22.0	3.8	9	63	2.3	13
Brunei and Malaysia	1.6	0.2	23	15	0.5	20
Indonesia	10.5	1.8	13	75	2.8	12
Australia	1.7	0.3	22	20	0.7	18
Europe						
Norway	6.0	1.0	17	2.5	0.1	24
United Kingdom	20.0	3.5	10	--	--	--
USSR	75.0	12.9	2	450	16.4	1
Africa						
Algeria	7.6	1.3	15	55	2.1	14
Libya	25.5	4.4	7	100	3.7	9
Nigeria	15.0	2.6	11	115	4.2	6
Gabon	1.5	0.3	24	10	0.3	22
Total	581.9	100.0		2,729	100.0	

Sources: Figures for China are from Tables 4.1 and 2.5; others are from the New York Times, January 6, 1975, pp. 10-11, which gives the original output figures in millions of barrels per day, here converted into millions of tons per year.

FIGURE 9.1

China and World Production of Crude Oil, 1973-75

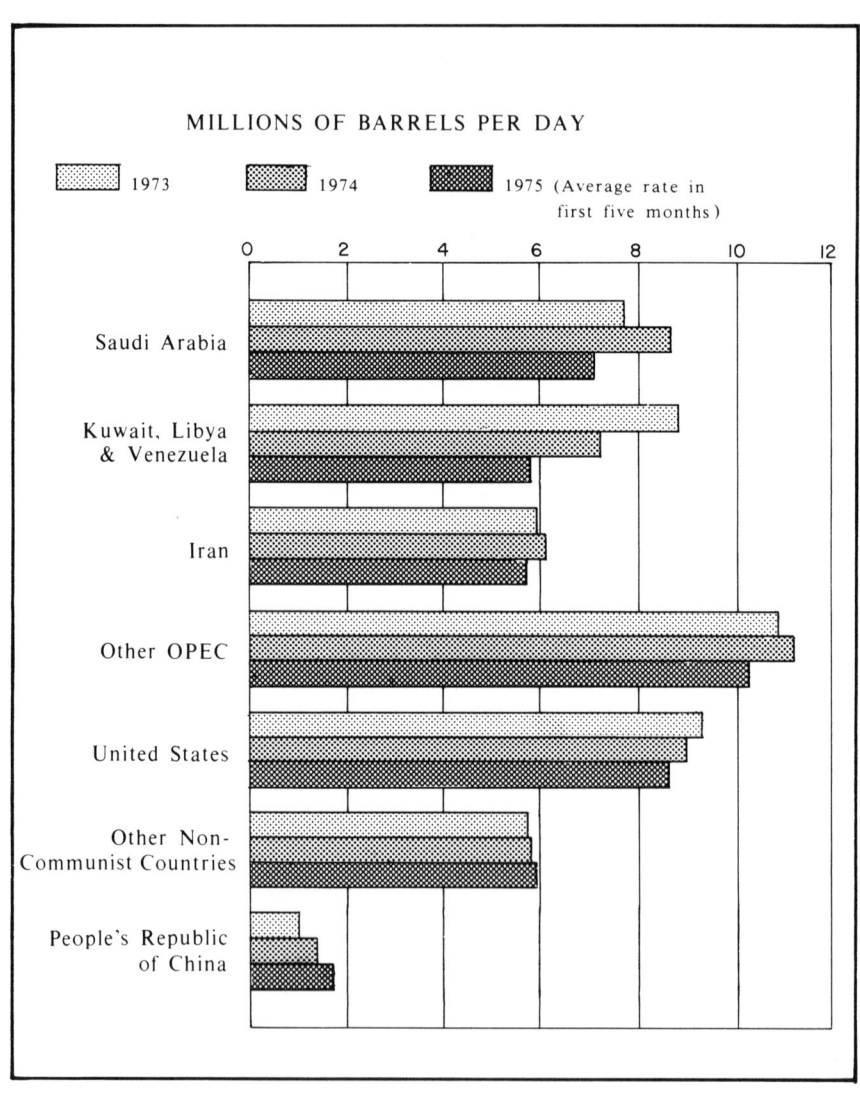

Source: New York Times, July 24, 1975.

in the United States, and 76.4 percent in Japan, as opposed to only 15 percent in China. In the years ahead other countries may lower their relative dependence on oil, and while China may increase her oil consumption, it is quite probable that her basic dependence upon coal is not likely to undergo any drastic changes in the near future.

TABLE 9.3

World Primary Energy Consumption,
by Regions and Sources, 1973
(in percent)

Regions and Countries	Oil	Natural Gas	Solid Fuels	Others
Latin America	74.0	14.9	6.0	5.1
Less developed countries in Eastern Hemisphere	57.7	6.3	34.3	1.7
China	15.0	15.0	69.0	1.0
Western Europe	63.3	9.7	23.6	3.3
Canada	51.8	26.8	12.0	9.4
United States	46.1	32.9	19.3	1.6
Japan	76.4	1.2	19.7	2.6

Sources: Figures for China are from Table 8.9; figures for other countries are from U.S. Department of State, The Effects of Rising Energy Costs on LDC Development, special report (Washington, D.C.: Government Printing Office, 1974), p. 2.

There are two major limitations on the Chinese petroleum industry, its extraction capacity and its transport facilities. The most promising oil resources lie on the continental shelf, from the Pohai Bay down to the South China Sea. Exploitation of this offshore reserve will require tremendous payments to the foreign producers of the Herculean drilling rigs and sophisticated equipment necessary to extract large quantities of oil from this area. On the other hand, extraction of the rich oil deposits located in the northwest oil fields (Karamai, Tsaidam, Tarim) will require the construction of many thousands of miles of pipeline in order to transport the crude to the exporting terminals on the eastern coast. Obviously the 5,000 kilometers (3,100 miles) of pipeline that were in operation as of 1974 represent a minuscule fraction of China's pipeline requirements; in comparison, in 1956 the United States had 78,000 miles of crude

FIGURE 9.2

World Energy Consumption by Region and Source, 1973

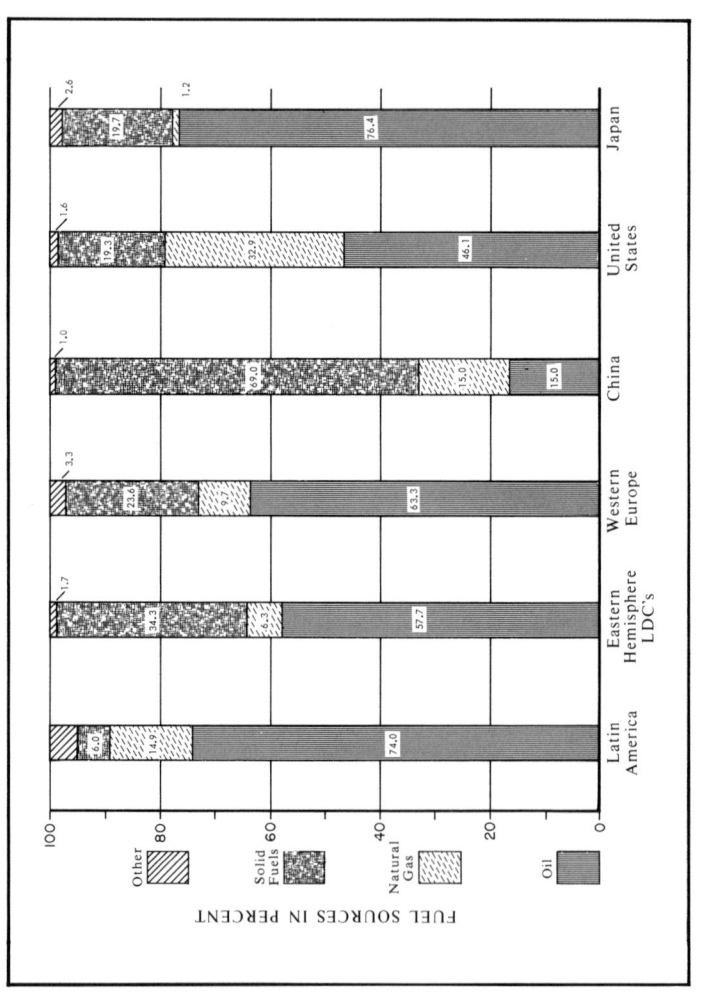

Source: Table 9.3

oil trunk pipeline, 39,000 miles of refined product pipeline, and
150,000 miles of gas trunk pipeline.[1] For China to become a major
oil exporter, a large-scale expansion of her pipeline system is a
prerequisite.

It is also impossible for China's feeble tanker fleet to transport tens of millions of tons of crude oil along the coast or convey it
to overseas markets. The estimated 400,000 tons of tankers operating in 1974 was only a tiny fraction of the world's tanker capacity.
China cannot simply expand her tanker fleet through large-scale
acquisitions from Japan and Western Europe because the Chinese
harbors lack the capacity to accommodate vessels over 100,000 tons.
It may take at least until the mid-1980s to modernize the major
Chinese harbors in order to accommodate tankers displacing more
than 100,000 tons.

Assuming that China attains her exporting goal of 100 million
tons by 1985, her relative share of the world oil trade will be little
more than 3 to 4 percent.

IMPLICATIONS FOR ECONOMIC DEVELOPMENT

China's relatively limited oil export potential for 1985 must
not be confused with the contribution of the domestic petroleum industry to her economic development. Until the 1970s the Chinese
economy was dominated by agriculture. The relationship between
agricultural output and other economic activities--not least among
them being capital accumulation--is extremely close in China. In
1957, in terms of net value-added, agriculture contributed more than
40 percent of the GNP. About 80 percent of the raw materials for
light industry and 50 percent for industry as a whole were derived
from agriculture. Approximately 50 percent of the government's
total fiscal revenue came directly or indirectly from agriculture.
Farm products in raw or processed form accounted for more than
70 percent of China's exports.[2] Past records corroborate that the
level of industrial output, foreign trade, and GNP were closely related to agricultural output until 1967. A somewhat changed relationship between agriculture and other sectors of the economy began
to emerge as petroleum output increased rapidly after 1967. During
the subsequent six years, despite the near-stagnation in food production (up only 3 percent over five years), industrial output rose 80 to
93 percent, the GNP gained 27 percent, and foreign trade went up
46 percent (see Table 8.1). This trend, if it continues, implies that
agriculture will probably have a diminishing influence on China's
general economic development.[3] In large part this conclusion is
based on the supposition that the increases in petroleum and chemical

industrial output, both using nonagricultural products as raw materials, render Chinese industry less dependent upon agriculture. Moreover, increased output of chemical fertilizers and other agricultural chemicals provides feedback from the industrial sector to agriculture, thus reversing the developmental strategy of the 1952-60 period, when agriculture was constantly squeezed to support heavy industry. Such a change should also help to improve agricultural productivity, which until the early 1970s was the Achilles' heel of the Chinese economy.

Within the industrial sector, the contribution of the growing petroleum industry to the modernization of China's industrial sector may be equally profound. The Chinese manufacturing industry is still relatively backward, overall, and underlying this backwardness is a lack of modern integrated plants. Most of the industrial facilities in China have been in operation for more than half a century. In 1965, of the Chinese machine-building output, 60 percent was reportedly turned out by manual labor, and half of the equipment was 30 to 50 years old.[4] Similar conditions also prevailed in other industries. Even those plants built with Soviet aid in the 1950s have become anachronisms. Replacement and modernization are imperative if China's industrial development is to advance rapidly. Because of her limited ability to pay, China has imported limited amounts of modern machinery and equipment from abroad. During the 22 years between 1952 and 1973, total overseas procurement of machinery and equipment amounted to only $9.2 billion.[5]

With the rising crude surplus expected in the late 1970s and early 1980s, China's ability to import foreign equipment and technical know-how will be substantially increased. Exporting the estimated total surplus of some 740 million tons of crude oil (5.4 billion barrels) between 1975 and 1985 (see Table 8.8) could generate a total revenue of $74 billion at the 1975 price of $100 per ton. This in turn would furnish China with enough foreign exchange to import at least five times the machinery and equipment she purchased abroad in the previous 22 years. The influx of this sizable quantity of equipment, representing the latest technology from Japan and the West, could potentially bring new vitality to the Chinese manufacturing industry.

It would appear that continued increases in China's petroleum output in the forthcoming decade will be a powerful factor in kindling a technological transformation in agriculture and industry, thereby accelerating the growth rate during the 1980s; yet for China to rival the Middle East and North America as a major oil producer or exporter may, if it ever happens, take an additional two or three decades.

NOTES

1. Shell International Oil Co., Petroleum Handbook (London: Shell, 1959), pp. 324-25.

2. Liu Jih-hsin, "On the Relationship between Agriculture and Heavy Industry," Ta-kung Pao, February 2, 1961, p. 3.

3. For details see Chu-yuan Cheng, "Economic Fluctuations in the PRC, 1949-1972," Current Scene 12, no. 7 (July 1974): 1-10.

4. Chu-yuan Cheng, The Machine-building Industry in Communist China (Chicago: Aldine-Atherton, 1971), p. 226.

5. Central Intelligence Agency, Republic of China: Foreign Trade in Machinery and Transportation Equipment since 1952 (Washington, D.C.: Central Intelligence Agency, January 1975), p. 1.

Books and Articles in Chinese

Chao Chun-chieh. "Current Problems in the Development of China's Shale Oil Industry." Shih-yu Kung-yeh T'ung-hsun (Peking), July 13, 1957.

_____. "Problems in Implementing the Policy of Diligence and Thriftiness in the Construction of the Petroleum Industry." Chi-hua Ching-chi [Planned Economy] (Peking), December 12, 1957, pp. 12-15.

Chao I-wen. Hsin-chung-kuo ti Kung-yeh [Industry in New China]. Peking: T'ung-chi Ch'u Pan-she, 1957.

Ch'en Cheng-siang. "Petroleum Resources and Their Development in China." Research Report no. 5, Geographical Research Center, Graduate School, The Chinese University of Hong Kong (Hong Kong), 1968.

Ch'en Ping-fan. Chung-kuo Chih Shih-yu [Petroleum in China]. Taipei: China Petroleum Company, 1962.

_____. Chung-kuo K'uang-ch'an Tsu-yuan [Mineral Resources of China]. 2 Vols. Taipei: Publications Committee on the Chinese Culture, 1954.

Ch'en Pu-ch'ing. "Natural Conditions and Resources on the Northern Slopes of the Tien Shan." Ti-li Chih-shih, May 1955, pp. 146-50.

_____. "Urumchi, the Garden of the Gobi Desert." Ti-li Chih-shih [Geographical Knowledge], December 1959, pp. 543-57.

Chien Yuan-heng. "An Analysis of Peking's Petroleum Production, Exportation and Transportation Capabilities." Chung-kung Yen-chiu [Studies on Chinese Communism] (Taipei), April 1975, pp. 98-103.

Chou Wen-fa. "Inspection Report on the Tsaidam Basin." Shih-yu K'an-t'an, no. 3 (February 7, 1958).

Chou Wen-lung. "Report before the Conference on Machinery Manufacture." Shih-yu K'an-t'an, no. 7 (April 7, 1959).

Chuan Wei. "A Critical Review on Recent Development of Petroleum Industry on Chinese Mainland." Ta-lu Ching-chi Yen-chiu [Studies of Mainland's Economy] 7, no. 1 (January 1975): 41-60.

Fang Chiang. Chung-kuo Ti-hsia Pao-ts'ang [China's Underground Wealth]. Hong Kong: Hsueh-sheng Book Company, 1957.

Fei-ch'ing Yen-chiu staff member. "Communist China's Petroleum Industry: An Eyewitness Report." Fei-ch'ing Yen-chiu 1 (Taipei), no. 9 (1967): 52-57.

Feng Sheng-wu. "The Past and Present of Lan-chou." Ti-li Chih-shih, June 1957, pp. 243-47.

Feng Ta-lin. Ti-i-ke Wu-nien Chi-hua Chung-kung-yeh ho yun shu-yeh ti Chi-pen Chien-she [The Basic Construction of Heavy Industry and Transportation during the First Five-Year Plan]. Peking, 1956.

Fu Chiao-chin. Shih-chieh Shih-yu Ti-li [Petroleum Geography of the World]. Peking: K'o-hsueh Ch'u-pan-she, 1959.

Ho Ko-jen. "The Development in Red China's Petroleum Industry." Fei-ch'ing Yueh-pao [Chinese Communism Monthly] (Taipei), March 1, 1968, pp. 52-65.

Ho Lan. Ti-i-ko Shih-yu-chi-ti Yu-men [Yumen--The First Petroleum Base]. Shanghai: People's Arts Publishing House, 1956.

Hsiao Chi-jung. Chung-kuo-ta-lu chih Neng-yuan yu Li-yung [Energy Resources and Their Utilization in Mainland China]. Taipei: Ministry of Economic Affairs, 1970.

Hsu P'ei-hsiu. "Geography of China's Oil Industry." Ti-li Chih-shih, October 1957, pp. 470-72.

K'ang Shih-en. "Address Before the Yumen Conference." Shih-yu K'an-t'an, August 15, 1958.

_____. "Conduct Another Forward Leap in Oil Prospecting with Revolutionary Vigor, Less Spending and More Work." Shih-yu K'an-t'an, March 5, 1958.

Ku Ching-hsin. "Shale Oil Industry of Our Country Achieved Rapid
    Progress." Hsin-hua Pan-yueh-k'an, June 11, 1959.

Kung-jen Jih-pao Editorial Department. Ta-ch'ing: Chung-kuo
    Kung-yeh-hua Ti Cheng-chueh Tao-lu [Tach'ing, The Correct
    Road of China's Industrialization]. Peking: Kung-jen Jih-pao
    she, 1966.

Li Chu-kuei. "How to Achieve More, Faster, Better, and Cheaper
    Construction for the Petroleum Industry." Shih-yu K'an-t'an,
    March 5, 1958, pp. 1-5.

Li Jen-chun. "The Petroleum Industry Can Leap Forward with
    Other Fellow Industries." Hsin-hua Pan-yueh-k'an, August
    16, 1958.

Li Ssu-kuang. "The Development of Geological Work During the
    Ten-Year Period Since the Founding of Our Country." Ti-chih
    Yueh-k'an [Geological Monthly], no. 10 (1959), pp. 17-20.

Li Tao-kwei. "Energy Crisis in the World and the Role that may be
    Played by the Chinese Communists." Ta-lu o ching-chi Yen-
    chin [Studies on the Economy of Mainland China] 5, no. 4 (1973):
    7-8.

Li Wen-yen. "Treasure Basin--Tsaidam." Ti-li Chih-shih, March
    1959, pp. 110-11.

Liu Fang. "The Road to the Development of Local Petroleum Indus-
    try." Shih-yu Lien-chih, June 6, 1958.

Lu Yun. "Conditions of Communist China's Electrical Equipment
    Industry." Fei-ch'ing Yen-chiu [Studies on Chinese Com-
    munism] (Taipei), November 1971, pp. 77-89.

People's Republic of China, State Planning Commission. "Problems
    to Consider in Implementing the 1958 Basic Construction Plan."
    Chi-hua Ching-chi, no. 2 (February 9, 1958), pp. 4-7.

People's Republic of China, State Statistical Bureau. "Basic Condi-
    tions in China's Mineral Survey and Prospecting Work."
    T'ung-chi Kung-tso, no. 5 (March 1957), pp. 31-32.

_____. "Communique on the Fulfillment of the First Five-Year
    Plan." Hsin-hua Pan-yueh-k'an, no. 8 (April 1959).

_____. "General Survey of the Distribution of the Centrally Allo-
cated Commodities in the Past Years." T'ung-chi Kung-tso,
no. 13 (July 1957), pp. 29-31.

Po I-po. "Report on the Fulfillment of the 1956 National Economic
Plan and the Draft Plan for 1957." Hsin-hua Pan-yueh-k'an,
July 14, 1957.

_____. "Strive for a New Victory in China's Industrial Production
and Construction." Hung-ch'i, February 1, 1961.

_____. "Ten Problems in the 1958 Economic Plan." Jen-min Jih-
pao, August 11, 1957.

Production-Economy Office, Shanghai Chemical Bureau. "Advance
along the Road of Overtaking and Surpassing Advanced World
Levels." Hua-hsueh Kung-yeh, no. 2 (1966), pp. 14-16.

Republic of China, Ministry of Economic Affairs. Ta-lu Neng-yuan
Chan-yeh Yen-chiu [Study on Energy Industry of Mainland
China]. Taipei: 1969.

_____, Research and Development Section. Ta-lu Shih-yu Kung-yeh
Kai-lan [An Overview of Petroleum Industry on the Chinese
Mainland]. Taipei: 1967.

Research Group on Mainland China's Petroleum and Chemical Industries. Ta-lu Shih-yu Kung-yeh Hsian-shih [Current Situation
of Petroleum Industry on the Chinese Mainland]. Taipei:
China Petroleum Company, 1968.

Shao Cheng-hung. "Cost Analysis for Chemical Products." Hus-
hsueh Kung-yeh [Chemical Industry], no. 14 (1963), pp. 22-24.

"Status of the Shallow Oil and Gas Formations and Possible Deposits
of Small Oil and Gas." Shih-yu K'an-t'an, July 17, 1958, pp.
43-45.

Sun Ching-chih, ed. Hua-pei Ching-chi Ti-li [An Economic Geography
of North China]. Peking: K'o-hsueh Ch'u-pan-she, 1957.

Sun Ching-chih, et al., eds. Hua-chung Ti-ch'u Ching-chi Ti-li
[An Economic Geography of Central China]. Peking: K'o-
hsueh Ch'u-pan-she, 1960.

_____. Hua-tung Ti-ch'u Ching-chi Ti-li [An Economic Geography of Eastern China]. Peking: K'o-hsueh Ch'u-pan-she, 1959.

Tang Ke-she. "Suggestions for the Geological Survey and Oil Prospecting Work in the Current Year." Shih-yu k'an-t'an, no. 7 (May 2, 1959).

Wang Chien. "Where is the Ta-ch'ing Oilfield." Tsu-kuo [China Monthly], Hong Kong, Union Research Institute, no. 5, August 1964.

Wang Hsin-san. Ti-i-kuo Wu-nien-chi-hua chung ti Jang-liao Kung-yeh [Fuel Industry in the First Five-Year Plan]. Peking: Chung-hua chuankuo K'o-shueh Chi-shu Pu-chi Hsie-hui, 1956.

Wo-kuo Kuo-ming Ching-chi Chuan-men Hsin-yueh-chin [The New Leap Forward of Our National Economy in All Fronts]. Hong Kong: San-lien Su-tien, 1967.

Wo-men Cheng Tsai Ch'ien-chin [We Are Marching Ahead]. Peking: Jen-min Ch'u-pan-she, 1972.

Wu Ch'uan-chien, et al., eds. Tung-pei Ti-ch'u Ching-chi Ti-li [An Economic Geography of Northeast China]. Peking: K'o-hsueh Ch'u-pan-she, 1959.

Yiu I-ming. "Sino-Japanese Trade: Current Development and Future Prospects." Ta-lu Ching-chi Yen-chiu, October 1974, pp. 41-62.

Books and Articles in Japanese

Chugoku Keizai Tekeyo [Abstracts on Chinese Economy]. Tokyo: Sekai Keizai Joho Sabisu, 1972.

Doi Akira. "The Production and Price of Petroleum from Chinese Ta-ch'ing Oil Field." Showa Dojin [Showa Coterie] (Tokyo), August 1970, pp. 18-22.

"The Expansion and Improvement of China's Port Facilities." Chugoku Kogyo Tsushin (Tokyo), February 1973, pp. 22-24.

Fuji Jahnalu, Chugoku Keizai Kenkyu-bu, Chugoku Keizai no Genjo to Tembo (Current Situation and Outlook of Chinese Economy), Tokyo, 1974.

Gaimusho-Keizai-gyoku, Keizai-togoka. "Petroleum--Resource Report of Continental China and Resource Policy of China." Tokyo: Ministry of Foreign Affairs, 1970.

Kagawa Kanichi. "Oil and Natural Gas Resources of China Continent." Tennen Gas [Natural Gas], March 1966, pp. 1-11.

_____, and Tanabe Yoshiyuki. "Oil and Natural Gas Resources of China Continent." Sekiyu-Gakkai-shi [Journal of Petroleum Society] 9, no. 10 (1969): 53-56.

Kambara, Tatsu. "Petroleum Industry in China." Sekiyu-no-Kaihatsu [Petroleum Exploration], April 1972, pp. 17-38.

_____. "Promising Petroleum Exploration in China." Chogoku Kogyo Tsushin [Bulletin of Chinese Industry] (Tokyo), July 2, 1972, pp. 17-22.

Kogyo-Gyi Jyutsu-in Chi-shi-tsu Cho-sa-sho [Institute of Geological Survey, Academy of Industrial Technology]. Collection of Latest Materials on China's Underground Resources. Tokyo, 1973.

Kudo Hirotada. "Chugoku Tairiku no Sekiyu Shigen" [Oil Resources of Continental China]. Asia Economic Institute Overseas Investment Material no. 9. Tokyo: Ajia Keizai Kenkyu Sho, 1966.

_____. "Energy Resources in the People's Republic of China." Japanese Science and Technology 13, no. 1 (January 1972): 33-48.

Ohno, Hideo. "Transition Period of China's Petroleum Industry." Chugoku Kogyo Tsushin [Bulletin of China's Industry] (Tokyo), February 1973, pp. 6-13.

Okubo Yasushi. Chugoku Gunji Kogyoryoku no Bunseki [An Analysis of the Military Power of China]. Tokyo: Asahi-shimbun Sha, 1968.

Onoe Etsuzo. Chugoko no Sangyo Ritchi ni Kansuru Kenkyu [A Study of Chinese Industrial Location]. Tokyo: Ajia Keizai Kenkyu Sho, 1970.

Ozuke Mitsugu. "The Present Situation and Problems of Chinese
    Crude Oil Exploitation." Maruzen Sekiyu Company Research
    Report no. 16. Tokyo, 1966.

"The Revival of Ta-ch'ing Oilfield." Nihhon Keizai Shimbun [Japan
    Economic Journal] (Tokyo), July 2, 1974.

Shinoda Nobuo. "China's Petroleum Industry--Development through
    Self-Reliance." Asahi Shimbun Research Department Report
    No. 124. Tokyo, 1966.

"The Situation of Current Development in Ta-ch'ing Oil Field."
    Naikaku-chosa-shitsu chosa-geppo [Cabinet Study Section
    Monthly Report] (Tokyo), May 1970, pp. 18-36.

"Ta-ch'ing Oil Field--A Giant Industrial and Agricultural Complex."
    Yomiuri Shimbun (Tokyo), July 2, 1974.

Takeishi Keinosuke. "What Does It Mean When China Exports Crude
    Oil to Japan?". Chugoku Kogyo Tsushin, February 1973, pp.
    2-5.

Togawa, Toru. "China's Petroleum Industry." Chugoku-Keizai-
    Kenkyu-Geppo [Study on Chinese Economy, Monthly]. (Tokyo),
    July 1973, pp. 1-52.

_____. "Chinese Oil and Petrochemical Industry." Kasen-Geppo
    [Chemical Fiber Monthly] (Tokyo), September 1971, pp. 40-48.

_____. "Petroleum Industry in China." Coke Digest (Tokyo), 1971,
    pp. 38-43.

Yoshida Hanehmon. "Petroleum of Communist China." Energy,
    April 1969, pp. 109-18.

                Books, Monographs, and Articles in English
                      or Translated into English

Abaydulla, Husayin. "The New Sinkiang." China Reconstructs 15,
    no. 1 (January 1966): 26-30.

Adelman, M. A. "Efficiency of Resource Use in Crude Petroleum."
    Southern Economic Journal 31 (1964): 101-20.

_____. The World Petroleum Market. Baltimore: The Johns Hopkins University Press, 1972.

American Petroleum Institute. Technical Report No. 2: Organiza-tion and Definitions for the Estimation of Reserve and Produc-tive Capacity of Crude Oil. Washington, D.C.: American Petroleum Institute, 1970.

"Annual Plans of Communist China--Results of Tsinghai Five-Year Plan." Joint Publications Research Service, no. 1844 (August 21, 1959), pp. 1-15.

Bell, Harold S. Petroleum Transportation Handbook. New York: McGraw-Hill, 1963.

Berizina, Y. I. "Toplivno-Energeticheskaya Baza Kitayskey Narodnoy Republiki" [Fuel and Power Base of the Chinese People's Republic]. Moscow: 1959, U.S. Joint Publications Research Service, Washington, D.C., no. 3784 (August 31, 1960), pp. 1-104.

Bradley, Paul G. The Economics of Crude Petroleum Production. Amsterdam: North-Holland Publishing Company, 1967.

Cain, Robert. Power Industry in Communist China. Hong Kong: Union Research Institute, 1969.

Campbell, Robert W. The Economics of Soviet Oil and Gas. Balti-more: Johns Hopkins Press, 1968.

Central Intelligence Agency. Foreign Trade in Machinery and Trans-portation Equipment since 1952. Washington, D.C.: Central Intelligence Agency, January 1975.

_____. People's Republic of China: International Trade Handbook. Washington, D.C.: Central Intelligence Agency, September 1974.

_____. Production of Machinery and Equipment in the People's Re-public of China. Washington, D.C.: Central Intelligence Agency, May 1975.

Chang, Chao. "Industry Comes to the Northwest." China Recon-structs 4, no. 4 (April 1955): 10-12.

Chang, Chun. "Peiping's Petroleum Industry: Growth and Future Development." Issues and Studies (Taipei) 10, no. 8 (May 1974): 41-56.

Chang, Kuei-sheng. "Geographical Bases for Industrial Development in Northwestern China." Economic Geography, 1963, pp. 341-50.

_____. Petroleum Resources and Production in Mainland China. Taipei: Institute of International Relations, 1963.

Chao, Kang. The Rate and Pattern of Industrial Production in Communist China. Ann Arbor: University of Michigan Press, 1965.

Chao, Yu-sheng. "The Tach'ing Oilfield." In Collected Documents of the First Sino-American Conference on Mainland China. Taipei: Institute of International Relations, 1971, pp. 795-821.

Chase Manhattan Bank. Capital Investments of the World Petroleum Industry. New York: the Bank, various years.

"Chemical Industry Develops in Northwest China." New China News Agency (Hsi-ning), May 13, 1959; Survey of China Mainland Press, no. 2017 (May 21, 1959), p. 13.

Ch'en Nai-ruenn. Chinese Economic Statistics. Chicago: Aldine Publishing Company, 1967.

Cheng Chu-yuan. China's Allocation of Fixed Capital Investment, 1952-1957. Ann Arbor: Center for Chinese Studies, The University of Michigan, 1974.

_____. "China's Industry: Advances and Dilemma." Current History 60, no. 361 (September 1971): 154-59.

_____. "China's Machine-building Industry." Current Scene 11, no. 7 (July 1973): 1-10.

_____. The Chinese Market under Communist Control. Hong Kong: Union Research Institute, 1955.

_____. Communist China's Economy 1949-1962. South Orange, N.J.: Seton Hall University Press, 1963.

_____. "Economic Fluctuations in the PRC, 1949-1972." Current Scene 12, no. 7 (July 1974): 1-15.

_____. Economic Relations Between Peking and Moscow. New York: Praeger, 1964.

_____. "The Effects of the Cultural Revolution on China's Machine-Building Industry." Current Scene 8, no. 1 (January 1970): 1-15.

_____. "Kansu." Encyclopaedia Brittanica. Chicago: Encyclopaedia Brittanica, 1974, pp. 387-92.

_____. The Machine-building Industry in Communist China. Chicago: Aldine-Atherton, 1971.

_____. Scientific and Engineering Manpower in Communist China. Washington, D.C.: National Science Foundation, 1966.

Cheng, Shih. A Glance at China's Economy. Peking: Foreign Languages Press, 1974.

Chien, Feng. "Northwest Survey." People's China, no. 3 (February 1, 1954), p. 15.

"China's New Frontier Trouble." Far Eastern Economic Review 42, no. 2 (October 10, 1963): 60-62.

"China's Oil Industry." Current Scene (Hong Kong) 40, no. 11 (November 1973): 21-23.

"China's Tsaidam Oil Fields Expanded." New China News Agency (Leng-hu), January 20, 1960; Survey of China Mainland Press, no. 2184, January 26, 1960, p. 21.

Chinese Communist Party. "Proposal of the Second Five-Year Plan." Eighth National Congress of the Chinese Communist Party. Peking: Foreign Languages Press, 1956.

"Chinese Oils Flow Up but Much Larger Gains Needed." Oil and Gas Journal, December 13, 1971, pp. 35-39.

Chu An-ping. "Karamai Oil." People's China 16 (August 16, 1956): 28-30.

"Communist China's Oil Industry." Far Eastern Economic Review
    26, no. 14 (April 2, 1959): 470-72.

Dudas, Gyula. "The Industrial Geography of the Chinese People's
    Republic." Foldraizi Kozlemenyek [Geographic Publications]
    9, no. 2 (1961); Joint Publications Research Service, no.
    11757 (January 2, 1962), pp. 4-45.

Dwyer, D. J. "The Coal Industry in Mainland China Since 1949."
    Geographical Journal 129, part 3 (September 1963): 329-38.

Ebel, Robert E. Communist Trade in Oil and Gas. New York:
    Praeger, 1970.

_____. The Petroleum Industry of the Soviet Union. Washington,
    D.C.: American Petroleum Institute, 1961.

Eckstein, Alexander. "China's Economic Growth and Foreign Trade."
    U.S.-China Business Review (Washington, D.C.), July-August
    1974, pp. 15-18.

_____, and Bruce Reynolds. "Sino-American Trade Prospects and
    Policy." American Economic Review, May 1974, pp. 294-99.

Economic Cooperation Administration. European Recovery Program,
    Petroleum and Petroleum Equipment Commodity Study. Wash-
    ington, D.C.: Government Printing Office, 1949.

Emerson, John Philip. Nonagricultural Employment in Mainland
    China, 1949-1958. U.S. Department of Commerce, Bureau of
    the Census. Washington, D.C.: Government Printing Office,
    1965.

First Five-Year Plan for Development of the National Economy of
    the People's Republic of China, 1953-1957. Peking: Foreign
    Languages Press, 1956.

"Five-Year Plan Implementation in Kansu." Joint Publications Re-
    search Service, no. 1592 (May 21, 1959), pp. 1-14.

Frick, Thomas C., ed. Petroleum Production Handbook. New York:
    McGraw-Hill, 1962.

"Fruit of Great Cultural Revolution: Tach'ing Is Five Times Its
    Former Self." Peking Review, no. 23 (1974), p. 16.

Gardner, S., and S. Hanke, eds. Essays in Petroleum Economics. Golden, Colo. Colorado School of Mines, 1967.

Georges, Michel. "Petroleum Exports Determined by Political Goals." Defense Nationale No. 7F--Problemes Politiques, Economiques, Scientifiques, Militaires, Paris, April 1974, reprinted in Joint Publication Research Service, no. 61,933 (May 8, 1974), pp. 1-10.

"Glaciers Explored." Far Eastern Economic Review 26, no. 22 (May 28, 1959): 742.

Gould, Sidney H., ed. Sciences in Communist China. Washington, D.C.: American Association for the Advancement of Science, 1961.

Griffin, James M. Capacity Measurement in Petroleum Refining. Lexington, Mass.: Heath, 1971.

Halh, Albert V. The Petrochemical Industry: Market and Economics. New York: McGraw-Hill, 1970.

Hartshorn, J. E. Politics and World Oil Economics. New York: Praeger, 1967.

Hassmann, Heinrich. Oil in the Soviet Union. Princeton, N.J.: Princeton University Press, 1953.

Heenan, Brian. "China's Petroleum Industry." Far Eastern Economic Review 49, no. 13 (September 23, 1965): 565-67.

Hirschman, Albert O. The Strategy of Economic Development. New Haven: Yale University Press, 1966.

Ho Ko-jen. "Peiping's Petroleum Industry." Issues and Studies 4, no. 1 (August 1968): 22-35.

Ho Ping-ti. "China's Resources Loom Large on World Stage: Huge Oil Deposits Could Weaken Arab Dominance." The Los Angeles Times, October 13, 1974, sec. 6, p. 1.

Hodges, John E., and Henry B. Steel. "An Investigation of the Problems of Cost Determination for the Discovery, Development and Production of Liquid Hydrocarbons and Natural Gas Resources." Rice Institute Pamphlet No. 46 (October 1959).

Hsia, Ronald. "Changes in the Industrial Sector Since 1953." Far
    Eastern Economic Review 27, no. 14 (October 1, 1959):
    536-39.

Hsieh, Chiao-min. Atlas of China. New York: McGraw-Hill, 1973.

Hsieh, Chia-yung. "Progress in Prospecting." China Reconstructs
    5, no. 1 (January 6, 1956): 5-7.

Hsueh, Pao-ting. "Soviet Aid to China's Industrialization." China
    Reconstructs 6, no. 11 (November 1957): 6-7.

Hua Ching-yuan. "New Achievements in China's Oil Industry."
    China Foreign Trade, no. 1 (1975), pp. 5-8.

Hung Yu-ch'iao. "An Appraisal of the Changing Fuel Structure on
    the China Mainland." Issues and Studies (Taipei), November
    1974.

Il'yin, A. I., and M. P. Voronichev. Zheleznodorozhnyy Transport
    Kiayskoy Narocnoy Respubliki [Railroad Transport of the
    Chinese People's Republic], Moscow, 1959. Joint Publica-
    tions Research Service, no. 3484 (July 6, 1960), pp. 1-151.

"Industrial Development of Sinkiang." Union Research Service 1,
    no. 12 (October 25, 1955): 1-9.

"Industry--China, 1949-1959, A Survey of Economic Change in
    China During One Decade of Communist Government." Far
    Eastern Economic Review 82, no. 14 (October 1, 1959): 536-49.

Jen, C. K. "My Impressions of the New China and Its Science and
    Technology." Eastern Horizon (Hong Kong) 12, no. 4 (1973):
    45-56.

Jones, P. H. M. "China Hustles Tsinghai." Far Eastern Economic
    Review 33, no. 6 (May 11, 1961): 250-54.

_____. "Sinkiang--China's Last Frontier." Far Eastern Economic
    Review 30, no. 5 (November 3, 1960): 200-205.

_____. "Sinkiang Unshackled." Far Eastern Economic Review 45,
    no. 5 (July 30, 1960): 197-98.

Jorgenson, Dale W. "U.S. Energy Policy and Economic Growth, 1975-2000." Paper presented at the Public Utilities Seminar, November 14, 1974, Indiana University, Bloomington, Indiana.

Juan, Vei Chow. "Mineral Resources of China." Economic Geology 41, no. 4, part 2, supplement (June-July 1946): 399-474.

Kalmykova, V. G., and I. Kh. Ovdiyenko. Severe-Zapadnyy Nitay: Geograficheskiy Ocherk [Geographical Survey of Northwest China]. Moscow: 1957; Joint Publications Research Service, nos. 1025-27N (December 12-15, 1958), pp. 1-157.

"Karamai--Newest and Biggest Oilfield." China Reconstructs 5, no. 11 (November 1956): 7-9.

Ku, Lei. "Tsaidam." China Reconstructs 6, no. 4 (April 1957): 2-5.

"Lanchow." Far Eastern Economic Review 28, no. 8 (May 5, 1960): 899.

"Lanchow Oil Refining Plant Rebuilt into a Chinese-Type Large Oil Refining Base." Union Research Service 45, no. 1 (October 10, 1966): 9-15.

Leszczycki, Stanislaw. "The Development of Geography in the People's Republic of China." Geography 43 (1963): 139-54.

Lewis, Christopher. "Outlook Bright for Oil, Coal." Far Eastern Economic Review, October 4, 1974, pp. 21-22.

Li, Choh-ming. "China's Industrial Development, 1958-1963." China Quarterly, no. 17 (January-March 1964), pp. 3-38.

_____. "Economic Development: The First Decade." China Quarterly, no. 1 (January-March 1960), pp. 35-50.

Li, Fu-chun. "Report of the First Five-Year Plan for the Development of the National Economy." People's China, August 15, 1955, supplement.

Ling, II. C. The Petroleum Industry of The People's Republic of China. Stanford, Calif.: Hoover Institution Press, 1975.

Liu, E. L. "Ho-si Corridor." Economic Geography 27 (January 1952): 51-56.

Liu, Kang. "Industry Comes to the National Minorities." People's
     China, no. 12 (June 16, 1956), pp. 15-18.

Liu, Tsai-hsing. "Changing the Economic Map of China." China
     Reconstructs 5, no. 5 (May 1956): 2-6.

_____. "China's Industries Spread Out." People's China, no. 9
     (May 1957), p. 15.

Ludlow, Nicholas. "China's Oil." U.S. China Business Review 1,
     no. 1 (January-February 1974): 20-31.

"Mainland China: Intensive Prospecting Drive." Far Eastern Eco-
     nomic Review 24, no. 2 (July 14, 1960): 85.

Maizels, Alfred. Industrial Growth and World Trade. Cambridge:
     Cambridge University Press, 1963.

Meyerhoff, A. A. "Development in Mainland China, 1949-1968."
     The American Association of Petroleum Geologists Bulletin
     54, no. 8 (August 1970): 1567-80.

National Petroleum Council. Impact of Oil Exports from the Soviet
     Bloc. Washington, D.C.: the Council, 1964.

"New Oilfield in Turfan Basin." New China News Agency (Urumchi),
     February 20, 1960. Survey of China Mainland Press, no. 1963
     (March 2, 1959), p. 32.

"New-Type Oilfield." Peking Review, no. 24 (1974), p. 23.

"Notes on the Tach'ing Oilfield." China Reconstructs, December
     1968, pp. 38-43.

Odell, Peter R. An Economic Geography of Oil. London: G. Bell
     and Sons, 1963.

"Oil Prospecting and Drilling Tasks Set for Northwest China." New
     China News Agency (Hsi-an), January 9, 1953; Survey of China
     Mainland Press, no. 489 (January 10-12, 1953), p. 29.

"Old Oilfields Make New Contributions." China Pictorial (Peking),
     no. 2 (1974), pp. 4-5.

Organization for Economic Cooperation and Development. Pipelines
     and Tankers. Paris: OECD, 1961.

People's Republic of China, State Statistical Bureau. Ten Great Years. Peking: Foreign Languages Press, 1960.

"Petroleum Industry." China News Analysis, no. 220 (March 14, 1958), pp. 1-5.

"Petroleum Industry, 1958-1961." China News Analysis, no. 406 (February 2, 1962), pp. 1-7.

"Petroleum Industry in Communist China." Ajia Kenkyu [Data for Asia Research] (Tokyo), no. 295 (November 21, 1961); Joint Publications Research Service, no. 13269 (March 29, 1962), pp. 1-18.

"Production of Artificial Petroleum Increased in Shensi Province." Union Research Service 17, no. 8 (October 27, 1959): 122-23.

Rasmussen, P. N. Studies in Intersectoral Relations. Amsterdam: North-Holland Publishing Co., 1956.

Rawski, Thomas. "Measuring China's Industrial Performance 1949-72." Paper presented at the Conference on Quantitative Measures of China's Economic Output, Washington, D.C., January 17-18, 1975.

"Recent Development in Sinkiang--I." Union Research Service 13, no. 15 (November 21, 1958): 207-27

"Recent Development in Sinkiang--II." Union Research Service 13, no. 16 (November 25, 1958): 228-46.

Rowson, Richard C. "China Tops Mideast, U.S. Oil Reserves." In The Bulletin, Overseas Press Club of America 30, no. 1 (January 1, 1975): 1.

Ryabukhin, G. Ye. Geologiya v Kitava [Geology of China]. Moscow: 1960; "Two Years Surveying and Development of Petroleum Reserves in Communist China." Joint Publications Research Service, no. 3672 (August 10, 1960), pp. 4-11.

"Selected Extracts on Petroleum Exploration in Communist China." Joint Publications Research Service, no. 1478 (April 13, 1959), pp. 3-7.

"Selected Translations and Extracts on the Machine Industry in Communist China." Joint Publications Research Service, no. 4032 (September 12, 1960), pp. 5-9.

Sell, George. The Petroleum Industry. London: Oxford University Press, 1963.

"Seven Artificial Petroleum Factories in Ninghsia." Union Research Service 17, no. 8 (October 27, 1959): 124.

Shabad, Theodore. China's Changing Map. New York: Praeger, 1972.

Shell International Oil Company. Petroleum Handbook. London: Shell, 1959.

Smil, Vaclav. "Communist China's Oil Exports: A Critical Evaluation." Issues and Studies, March 1975, pp. 74-77.

_____. "Energy in the PRC." Current Scene 13, no. 2 (February 1975): 1-10.

"Ta-ch'ing Oil Field." China Pictorial (Peking), special issue, July 1966.

"Tarim Mining Bureau Prospecting Petroleum." Union Research Service 17, no. 4 (October 13, 1959): 66.

Tien, Chih-chun. "Petroleum Output Shows Marked Rise." China Reconstructs 12, no. 4 (April 1963): 6-7.

"Today's Karamai." Union Research Service 17, no. 4 (October 13, 1959): 56-59.

Tregear, T. H. An Economic Geography of China. London: Butterworths, 1970.

_____. A Geography of China. Chicago: Aldine Publishing Company, 1965.

"Tsinghai Oil Center Develops Many-sided Industries." New China News Agency (Hsi-ning), August 2, 1960; Survey of China Mainland Press, no. 2313 (August 9, 1960), p. 25.

Uchida, Genko. "Technology in China." Scientific American, November 1966.

United Nations, Department of Economic and Social Affairs. Techniques of Petroleum Development. New York: the UN, 1964.

"Up-to-Date Oil Cracking Plant Built by Sino-Soviet Joint Stock
Petroleum Company in Sinkiang." New China News Agency
(Urumchi), September 25, 1953; Survey of China Mainland
Press, no. 659 (September 29, 1953), pp. 44-45.

U.S. Congress, Joint Economic Committee. China: A Reassessment of the Economy. 94th Cong., 1st sess. Washington,
D.C.: Government Printing Office, 1975.

_____. An Economic Profile of Mainland China. 92nd Cong.
Washington, D.C.: Government Printing Office, 1972.

_____. People's Republic of China: An Economic Assessment.
92nd Cong., 2nd sess. Washington, D.C.: Government
Printing Office, 1972.

"USSR Aid for Lanchou Oil Refinery." Far Eastern Economic Review 27, no. 21 (November 19, 1959): 805-808.

"Vast Northwest to be Transformed into a Great Industrial Base."
New China News Agency (Hsi-an), April 13, 1953; Survey of
China Mainland Press, no. 552 (April 16, 1953), p. 19.

"Vast Population Movement to Communist China's Wasteland."
Union Research Service 6, no. 16 (February 22, 1975): 217-28.

Wageman, J. M., W. C. Thomas, and K. O. Emery. "Structural
Framework of East China Sea and Yellow Sea." American
Association of Petroleum Geologists Bulletin 54, no. 9 (1970):
1611-43.

Wallace, F. Lovejoy, and Paul T. Homan, with Charles O. Gavin.
Cost Analysis in the Petroleum Industry. Dallas: Southern
Methodist University, 1963.

Wang, K. P. "Natural Resources and Their Utilization." In China:
A Handbook, ed. Wu Yuan-li. New York: Praeger, 1973,
pp. 71-84.

_____. "The Mineral Resource Base of Communist China." In An
Economic Profile of Mainland China. Washington, D.C.: Government Printing Office, 1967, pp. 167-95.

_____. "A Review of Mining and Metallurgy." In Science in Com-
    munist China. Washington, D.C.: American Association for
    the Advancement of Science, 1961, pp. 687-738.

_____. "Rich Mineral Resources Spur Communist China's Bid for
    Industrial Power." Mineral Trade Notes, special supplement
    no. 59. Washington, D.C.: Government Printing Office,
    1960, pp. 1-35.

Washington, University of, Far Eastern and Russian Institute. A
    Regional Handbook of Northwest China. New Haven: Human
    Relations Area Files, 1956.

Weller, J. M. "Petroleum Possibilities of Reserve-Basin of
    Szechuan Province, China." American Association of Petro-
    leum Geologists Bulletin 28, no. 10 (1944): 1430-39.

Whitson, William W., ed. Doing Business with China: American
    Trade Opportunities in the 1970s. New York: Praeger, 1974.

Wu, Yuan-li, ed. China: A Handbook. New York: Praeger, 1973.

_____. Economic Development and Use of Energy Resources in
    Communist China. New York: Praeger, 1963.

Yeh, K. C. Communist China's Petroleum Situation. Santa Monica,
    Calif.: Rand Corporation, 1962.

Yu, Frederick. "Changes in Chinghai." Far Eastern Economic Re-
    view 30, no. 1 (October 6, 1960): 12-15.

Yung, Lung-kwei. "How China Builds Up Her Economy." China
    Reconstructs, no. 12 (December 1963), pp. 2-5.

Periodicals Published in People's Republic of China

Chi-hsieh Kung-ch'eng Hsueh-pao [Journal of Mechanical Engineer-
    ing], quarterly, Peking.

Chi-hsieh Kung-jen [Machinery Workers], monthly, Peking.

Chi-hsieh Kung-yeh [Machine-building Industry], organ of the First
    Ministry of Machine-building Industry, semimonthly, 1953-58;
    changed between January 1959 and June 1960 to Chi-hsieh

Kung-yeh Chou-pao [Machine-building Weekly]; suspended between July 1960 and December 1961; resumed publication as Chi-hsieh Kung-yeh in January 1962; suspended again in August 1966.

Chi-hua Ching-chi [Planned Economy], State Economic Commission and State Planning Commission official publication; published monthly between 1954 and 1958 and merged with T'ung-chi Kung-tso to become Chi-hua yu T'ung-chi in January 1959; published until the end of 1960.

China Pictorial (in English), monthly, Peking.

China Reconstructs (in English), monthly, Peking.

Ching-chi Chou-pao [Economic Weekly], Shanghai.

Ching-chi Yen-chiu [Economic Research], journal of the Economic Research Institute of the Chinese Academy of Sciences, monthly, Peking.

Ch'i-ts'ai Hang-ch'ing [Market Prices of Equipment and Materials], published by the Ministry of Heavy Industry, semimonthly, Peking.

Chung-hsing Chi-hsieh [Heavy Machinery], organ of the Bureau of Heavy Machinery, the First Ministry of Machine-building Industry, monthly, Peking.

Chung-kung-yeh T'ung-hsun [Bulletin of Heavy Industry], organ of the Ministry of Heavy Industry, monthly, Peking.

Chung-kuo Kung-jen [China's Workers], monthly, Peking.

Chung-kuo Kung-yeh [Industry of China], monthly, Shanghai.

Chung-kuo Nung-yeh Chi-hsieh [China's Agricultural Machinery], monthly, Peking.

Chung-kuo Tui-wai Mao-i [China's Foreign Trade], organ in Chinese and English of the Ministry of Foreign Trade, monthly, Peking.

Hsin-hua Yueh-pao [New China Monthly] and Hsin-hua Pan-yueh-k'an [New China Semimonthly], official People's Republic of China publications; sources of important government directives and selected articles on political, economic, cultural, and international affairs; monthly as <u>Hsin-hua Yueh-pao</u> from 1949 to 1955 and since 1961; semimonthly as <u>Hsin-hua Pan-yueh-k'an</u> from 1956 to 1960, Peking.

<u>Hsueh-hsi</u> [Study], theoretical journal for Communist cadres, monthly prior to 1956; semimonthly after 1956; suspended in 1958, Peking.

<u>Hua-hsueh Kung-yeh</u> [Chemical Industry], organ of the Ministry of Chemical Industry, monthly, Peking.

<u>Hung-ch'i</u> [Red Flag], published by Chinese Communist Party Central Committee, monthly since June 1958; irregularly since November 1967, Peking.

<u>Jen-min Tien-yeh</u> [People's Power Industry], monthly, Peking.

<u>Mei-t'an Kung-yeh</u> [Coal Industry], organ of the Ministry of Coal Industry, semimonthly, Peking.

<u>Nung-yeh Chi-hsieh Chi-shu</u> [Agricultural Machinery Techniques], monthly, Peking.

<u>Peking Review</u> [in English], weekly, Peking.

<u>People's China</u> [in English], semimonthly, suspended in January 1958 but continued in Japanese and Spanish.

<u>Shang-hai Chi-hsieh</u> [Shanghai Machinery], monthly, Shanghai.

<u>Shih-shih Shou-ts'e</u> [Current Events], semimonthly, Peking.

<u>Shih-yu K'an-t'an</u> [Petroleum Prospecting], semimonthly, Peking, specializing in geological prospecting and exploration.

<u>Shih-yu K'uai-pao</u> [Petroleum Express], Peking, reporting new technology in foreign countries.

<u>Shih-yu Kung-yeh T'ung-hsun</u> [Bulletin of the Petroleum Industry], contains major documents and technological information relevant to the Chinese petroleum industry, semimonthly, Peking.

<u>Shih-yu Lien-chih</u> [Petroleum Refining], focusing on oil production and refining, semimonthly, Peking.

Shui-li Yu Tien-li [Water Conservation and Electrical Power], semi-
    monthly, Peking.

Ti-li Chih-shih [Geographical Knowledge], monthly, Peking.

Ts'ai-cheng [Public Finance], organ of the Ministry of Public
    Finance, monthly, Peking.

T'ung-chi Kung-tso [Statistical Work], State Statistical Bureau offi-
    cial publication, monthly, Peking, 1956-59.

T'ung-chi Kung-tso T'ung-hsun [Statistical Work Bulletin], predeces-
    sor of T'ung-chi Kung-tso, semimonthly, Peking, 1954-56.

T'ung-chi Yen-chiu [Statistical Research], organ of the State Statis-
    tical Bureau, monthly, 1958, Peking.

Newspapers and News Agencies of
People's Republic of China

Ch'ang-chiang Jih-pao [Yangtze Daily], Hankow, Hupeh Province.

Chieh-fang Jih-pao [Liberation Daily], Shanghai.

Chung-kuo Ch'ing-nien Pao [China Youth Daily], Peking.

Chung-kuo Hsin-wen [China News Service], Canton.

Hsin-wen Jih-pao [News Daily], Shanghai.

Jen-min Jih-pao [People's Daily], Peking.

Kirin Jih-pao [Kirin Daily], Ch'angch'un, Kirin Province.

Kuang-ming Jih-pao [Enlightenment Daily], Peking.

Kung-jen Jih-pao [Worker's Daily], Peking.

Liaoning Jih-pao [Liaoning Daily], Shen yang, Liaoning Province.

Nan-fang Jih-pao [Southern Daily], Canton, Kwangtung Province.

New China News Agency, Peking.

Pei-ching Jih-pao [Peking Daily], Peking.

Sinkiang Jih-pao [Sinkiang Daily], Urumchi, Sinkiang.

Szechwan Jih-pao [Szechwan Daily], Ch'engtu, Szechwan Province.

Ta-kung Pao [Impartial Daily], Shanghai, Tientsin, Peking, and
    Hong Kong.

Tientsin Jih-pao [Tientsin Daily], Tientsin, Hopeh Province.

Tsingtao Jih-pao [Tsingtao Daily], Tsingtao, Shangtung Province.

Wen-hui Pao [Wen-hui Daily], Shanghai and Hong Kong.

Broadcasts and Translations in English Originating
Outside the People's Republic of China

China News Summary, Hong Kong.

Current Background, Hong Kong, U.S. Consulate General.

Extracts from China Mainland Magazines, Hong Kong, U.S. Con-
    sulate General.

Extracts from China Mainland Publications, Hong Kong, U.S. Con-
    sulate General.

Foreign Broadcast Information Service--Daily Report, Washington,
    D.C.

News from Chinese Provincial Radio Stations, Hong Kong.

Reports on China, Joint Publications Research Service, Washington,
    D.C.

Selections from China Mainland Magazines, Hong Kong, U.S. Con-
    sulate General.

Surveys of China Mainland Press, Hong Kong, U.S. Consulate
    General.

Union Research Service, Hong Kong.

CHU-YUAN CHENG is Professor of Economics at Ball State University, Muncie, Indiana. He also served as a consultant to the National Science Foundation during 1966-75.

A former director of the Research Department of Union Research Institute in Hong Kong and a Senior Research Economist at the University of Michigan, Professor Cheng has been a keen student of China's economy since the founding of the Chinese People's Republic in 1949, and has authored 15 books and monographs on the economic development of Mainland China. His major works published in this country include Communist China's Economy 1949-1962, Economic Relations between Peking and Moscow, Scientific and Engineering Manpower in Communist China, The Machine-building Industry in Communist China, and China's Allocation of Fixed Capital Investment. He has also contributed numerous articles and reviews to periodicals in the United States, Hong Kong, Tokyo, and London.

Professor Cheng received his B.A. from the National Chengchih University in Nanking. He came to the United States in 1959 and subsequently received his M.A. and Ph.D. in economics from Georgetown University, Washington, D.C.

CHINA'S CHANGING ROLE IN THE WORLD ECONOMY
edited by Bryant Garth and the Editors of
the Stanford Journal of International Studies

THE SOVIET ENERGY BALANCE: Natural Gas, Other
Fossil Fuels, and Alternative Power Sources
Iain F. Elliot

ARAB OIL: Impact on Arab Nations and Global
Implications
edited by Naiem A. Sherbiny
and Mark A. Tessler

DEVELOPMENT OF THE IRANIAN OIL INDUSTRY:
International and Domestic Aspects
Fereidun Fesharaki

*CHEMICAL AND PETRO-CHEMICAL INDUSTRIES
OF RUSSIA AND EASTERN EUROPE, 1960-1980
Cecil Rajana

ECONOMIC GROWTH AND EMPLOYMENT PROBLEMS
IN VENEZUELA: An Analysis of an Oil-Based Economy
Mostafa F. Hassan

*For sale in the United States and Philippines only.